Geralda Gillian Silva Sena
Flavia de Paula
Maria C. P. Batitucci

Human Longevity

Geralda Gillian Silva Sena
Flavia de Paula
Maria C. P. Batitucci

Human Longevity

From scientific knowledge to strategies for healthy ageing

ScienciaScripts

Imprint

Cover image: www.ingimage.com

This book is a translation from the original published under ISBN 978-613-9-67001-7.

Publisher:
Sciencia Scripts
is a trademark of
Dodo Books Indian Ocean Ltd. and OmniScriptum S.R.L publishing group

120 High Road, East Finchley, London, N2 9ED, United Kingdom
Str. Armeneasca 28/1, office 1, Chisinau MD-2012, Republic of Moldova, Europe
Printed at: see last page
ISBN: 978-620-8-10554-9

SUMMARY

HUMAN LONGEVITY

From scientific knowledge to strategies for promoting healthy ageing

About the Author

Geralda Gillian Silva-Sena has a degree in Nutrition (UFOP). Master in Plant Biology (UFES) and PhD in Biotechnology (UFES/RENORBIO). Two consecutive winners of the Longevity Research Award, Geriatrics category (2016 and 2017), Sao Paulo. Adjunct Professor at the Federal University of Espirito Santo.

About the co-authors

Flavia de Paula has a degree in Biological Sciences (UFES) and a PhD in Biology/Genetics (USP). She is an Associate Professor at the Federal University of Espirito Santo and advisor to the Biotechnology/Renorbio/CCS/UFES Postgraduate Program.

Maria do Carmo Pimentel Batitucci has a Bachelor's degree (emphasis in Genetics) and a Licentiate degree in Biological Sciences from the Federal University of Minas Gerais. She has a Master's degree in Genetics and Plant Breeding (UFV) and a PhD in Physiological Sciences (UFES). Full Professor at the Federal University of Espirito Santo and of the Postgraduate Programs in Plant Biology (UFES) and Pharmaceutical Sciences (UFES).

I dedicate it to my loves, Marcelo and our daughter, Lais.

"Those wrinkles that furrow the old woman's brow; that recumbent and tremulous walk, those short words, dense with experience and wisdom; that sweet look of a girl and a woman at the same time, but kinder than either, is a beauty that we don't know" Chiara Lubich.

"Everyone's path is slippery... but falling doesn't hurt too much - you get up, you climb, you come back!... Life is like that: it heats up and cools down, it tightens up and then loosens up, it quiets down and then quiets down. What she wants from us is courage." Guimaràes Rosa.

ACKNOWLEDGMENTS

To GOD, who, in his silence, welcomed me in times of difficulty, helped me overcome obstacles and start again...

To my family, for their support and for cheering me on...

To my companion of all hours, Marcelo, for his constant presence during all these years... Thank you for supporting me in my uncertainties, expectations, failures and for cheering me on to my victories! Only the two of us know how difficult it was to achieve this victory...

To Professor Aneli de Melo Barbosa and Dr. Robert Dekker, for the opportunity to learn about the botriosferan molecule... it was enriching!

To my friends and teachers that academic life has given me, for their support whenever necessary...

1 INTRODUCTION

In recent decades, there has been a worldwide increase in the elderly population due to a reduction in fertility and mortality rates. This increase in life expectancy, both in developed and developing countries, is related to improvements in basic sanitation, socio-economic conditions, lifestyle, as well as advances in health research (KINSELLA; HE, 2009).

However, in Brazil and other developing countries, the increase in the elderly population has occurred without the proper implementation of social, economic and health policies, in a context of fragile institutions and services for this vulnerable group (NOGUEIRA et al., 2008). Thus, there is a need to adapt these policies, as well as to include the elderly in research and increase studies on ageing, longevity and quality of life.

Ageing compromises individuals' ability to respond to environmental stress and to maintain homeostasis (ANSELMI et al., 2009). As a result of these peculiar changes, Chronic Degenerative Diseases (CDDs) such as hypertension, cardiovascular diseases, diabetes, obesity, dyslipidemia, cancer, rheumatological diseases and neurodegenerative diseases such as Alzheimer's and Parkinson's often appear.

Thus, studies on ageing in the last two decades have sought to establish models that explain the relationship between the onset of disease and age, but also those relating to positive phenotypes such as longevity and healthy ageing (DAVINELLI; WILCOXX; SCAPAGNINI, 2012; FALLIN; MATTEINI, 2009).

Human longevity is a complex phenomenon that can be influenced by the interaction of multiple environmental, behavioral and genetic factors (BROOKS-WILSON, 2013). The study of its possible predictors or markers is one of the most pressing needs/priorities, since it can enable research to advance in the prevention and reduction of diseases that accompany the length of life.

The way in which genetics can be a determining factor in longevity has provoked great interest among researchers, with approximately 25% of the variation in life expectancy in an adult being attributable to genetic differences. In addition, this influence seems to increase after the age of sixty, becoming more pronounced at the age of eighty-five or older (HERSKIND et al., 1996; HJELMBORG et al., 2006; PERLS et al., 1998).

Genes such as *FOXO3 (Forkhead box O3), SOD2 (Superoxide dismutase 2), APOE (Apoliprotein E) and SIRT1 (Silent Information Regulator Type 1)* have been described as lifespan modulators and are involved in lipid and carbohydrate metabolism pathways, the response to oxidative stress and susceptibility to the development of diseases that accompany aging (BARZILAI et al., 2012; DATO et al., 2013).

The decrease in the activity of antioxidant agents, accompanied by an excessive production of reactive oxygen species (ROS), which characterizes the process of oxidative stress, has also been linked to ageing and its morbidities. ROS, which are extremely unstable, can react with proteins, deoxyribonucleic acid (DNA) and lipids, altering cellular homeostasis (HARMAN, 1991; MATÉS et al., 1999). Lipids are easily accessible endogenous targets and are quickly affected by their effects, which favors the formation of secondary compounds such as malondialdehyde, widely used as a marker of the process (DE ZWART et al., 1999; ESTERBAUER; SCHAUR; ZOLLER, 1991).

Oxidative damage to DNA also appears to play an important role both in the process of senescence and the onset of chronic degenerative diseases, as well as in healthy ageing, and its levels in biological fluids such as plasma and tissues have been considered reliable biomarkers. Among the various proposals, the Comet Assay is one of the simplest and most widespread techniques for genotoxicological studies in higher organisms (NERI et al., 2015).

Investigating these critical determinants of lifespan in populations that have not yet been studied can improve scientific knowledge and help develop strategies related to biogerontology. The prevention and control of DCDs are also fundamental to avoiding adverse consequences for quality of life and losses in economically active life years in developing countries (BRASIL, 2011).

In this sense, in line with the Pact for Health and the National Policy for the Health of the Elderly, the Ministry of Health promulgated the Strategic Action Plan for Tackling Chronic-Degenerative Diseases in Brazil, from 2011 to 2022. This plan aims to prepare Brazil to face and stop these diseases through integrated, sustainable and evidence-based actions, considering health promotion, comprehensive care for the elderly and food as a modifiable risk factor for CDDs as guidelines (BRASIL, 2011; BRASIL, 2006a;b).

Davinelli, Wilcoxx and Scapagnini (2012) also described that the study of aging mechanisms in the most varied groups of living beings points to diet as an important modulator of the process, both in aging *per se* and in longevity itself, extending or reducing it.

As health promotion actions, the use of glucans, biopolymers with potential biotechnological application, has been highlighted because they can help prevent and control DCDs (SYNYTSYA; NOVAK, 2013). Fungal exopolysaccharides, such as (1→3^→6)-β-D-glucan botriosferan, have been extensively explored for their possible benefits in the prevention and treatment of diseases that accompany ageing, as well as longevity (KAGIMURA et al., 2015).

In view of the above, the investigation of genetic factors and their association with human longevity, as well as the influence of environmental characteristics, oxidative and genomic damage on lifespan, conducted in elderly populations that have not yet been studied, can improve scientific knowledge and establish more effective public monitoring/care strategies, whether preventive or diagnostic, according to the genetic profile of each patient.

In addition, studies aimed at evaluating the potential biotechnological application of botriosferan can help establish therapeutic and preventive interventions for various age-related diseases that cause significant mortality and morbidity rates and burden the public health system.

Chapter 1 Ageing: a new world and Brazilian scenario

Population ageing, the most notable demographic change of the 21st century, has been widely recognized as a worldwide phenomenon, with a field of relevant discussions.

According to the World Health Organization (WHO), from a chronological point of view, the elderly are individuals aged 65 or over living in developed countries and 60 or over in developing countries. This classification takes into account economic, social, political, cultural, environmental and quality of life factors, which determine the age patterns mentioned above (CAMARANO; KANSO; LEITÂO E MELLO, 2004).

The elderly population already exceeds 900 million individuals, which corresponds to 12.3% of the world's population, and is estimated to be close to 22.0% by 2050 (WHO, 2015). This is due to the reduction in mortality and birth rates, due to improvements in basic sanitation, socioeconomic conditions, lifestyle, as well as advances in health research, which are more evident in developing countries (KINSELLA; HE, 2009).

In Brazil, it is possible to observe the replacement, in the population pyramid, of the broad base by a rapidly and considerably widening summit, indicating, proportionally, more elderly people and fewer children and young people in the population (Figure 1) (IBGE, 2008).

This population makes up 12.6% of the current population, with projections of 16.2% by 2025. Life expectancy at birth for Brazilians has risen from 69.8 to

73.8 years between 2000 and 2010. In addition, by 2050, our country will rank sixth in the world in terms of the number of elderly people, with more than 13 million over the age of 80 (IBGE, 2008; IBGE, 2013a).

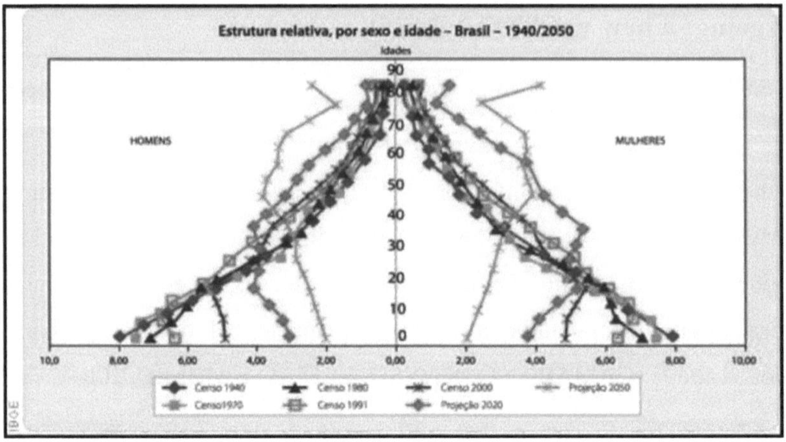

Figure 1: Relative *age* structure, by age and sex, in millions of people, in Brazil - 1940 to 2050 (Adapted from IBGE, 2008).

In the period from 2012 to 2022, a growth rate of more than 4.0% is expected, resulting in more than 1.0 million elderly people every year (IBGE, 2015). The ratio between the number of elderly people for every group of 100 children aged 0 to 14 is also changing: in 2010, there were 26.7 elderly people for every 100 children, but by 2050 the ratio is expected to be 6.5 times higher (172.7), with a significant concentration in the South and Southeast regions (IBGE, 2008).

In Espírito Santo, similarly, the elderly represent 12.7% of the population, with life expectancy projected to increase from 70.4 to 81.2 years between 2000 and 2030 (IBGE, 2013b).

According to Camarano, Kanso, Leitao and Mello (2004), it is also important to note that among the elderly, the number of those known as "very elderly" is also growing. This increase in the proportion of very old people is due to the reduction in the mortality rate at older ages.

In addition, scientific evidence shows that the group of individuals aged 60 and over cannot be considered uniform, due to the metabolic, physiological and socioeconomic differences observed between the elderly over 85 and those who are younger (BALTES; SMITH, 2006; JACOB FILHO, 2006; JOHNSON, 2005).

In this sense, the WHO has suggested a subdivision for this group: *young-old*, aged between 60 and 69; *old-old, aged* between 70 and 84; and *oldest-old*, aged 85 or over (De ONIS; HABICHT, 1996; WHO, 2004).

In Brazil, studies on ageing began in the 1980s, with the onset of the considerable age changes described above (LIMA; MENEZES, 2011). However, the production of knowledge is still incipient, with most research concentrating on describing clinical cases or case reports of the multiple diseases that affect this group, such as chronic-degenerative diseases and geriatric syndromes, as well as therapeutic or surgical procedures.

At the same time, as part of this thematic universe, pioneering research has been or is being carried out, of an epidemiological nature, but which seeks to elucidate possible cellular and molecular mechanisms, concentrated in the South and Southeast regions.

In the south of Brazil, programs such as Genesis (*Gene Invironmental Systems Interactions on Susceffull* Aging) and Veranópolis (Studies in Aging, Longevity and Quality of Life) have contributed to the development of research into genetic-environmental interactions in healthy aging of elderly people aged 80 and over. These investigations are taking place in the cities of Gravatai and Veranópolis, Rio Grande do Sul, which have a higher average life expectancy than the rest of the country (CRUZ et al., 2005; DA CRUZ et al., 2004). In the Southeast, the population-based BAMBUi-MG project set out to identify predictors of adverse health events in the elderly. Also in this region, the SABE (Health, Well-being and Ageing) project, coordinated by the Pan American Health Organization, aimed, in its first phase, to collect information on the health conditions of elderly people over 60 in urban areas of Latin American metropolises, including Sao Paulo (GIACOMIN et al., 2005; LEBRAO; LAURENTI; 2005).

Although ageing is one of humanity's great achievements, it is also a challenge for society and the academic community in different countries, especially those in the developing world. In Brazil, unlike in developed countries, the elderly population, with a fragile phenotype and a significant number of diseases, has been increasing in a context of social inequalities, economic instability and health services (PALLONI et al., 2002).

Thus, there is an undeniable need to reorganize socio-economic, academic-scientific and health policy structures to meet the specific demands of this group, with the establishment of more effective preventive, therapeutic and diagnostic interventions in geriatrics that improve quality of life and provide healthy ageing, with relevant impacts in terms of public health.

Chapter 2 Human Ageing and Postulated Theories

Aging or senescence is associated with two processes, progressive cellular degeneration and loss of regenerative capacity, and is defined as a progressive, functional and organic decline, influenced by molecular, cellular and systemic events, with alterations in homeostasis and, consequently, increased susceptibility to disease and death (SCHUMACHER; GARINIS; HOEIJMAKERS, 2008).

The mechanisms involved in the aging process have not yet been fully elucidated, but it is known that it is one of the main risk factors for the development of chronic degenerative diseases such as hypertension, cardiovascular diseases, diabetes, obesity, dyslipidemia, cancer, rheumatological diseases and neurodegenerative diseases such as Alzheimer's and Parkinson's (WHEELER; KIM, 2011). Due to the considerable increase in these diseases over the course of life, understanding the biological processes related to the pathophysiology of this event is of paramount importance.

Since ancient times, there have been discussions about the limits between the natural aging process and the onset of disease or senility. In the 7th century BC, the Greek physician Pithagoras already divided the human lifespan into cycles, the last of which, between 63 and 81 years, was the *senium* phase, marked by decline and incapacity of the organism (HALPERT, 1983).

However, only three decades ago, with the isolation of the first "long-lived" strains of the free-living nematode *Caenorhabditis elegans* by Klass (1983) and the beginning of possible analogies with the human life cycle, studies on aging made unprecedented progress.

A considerable number of theories have been put forward to explain ageing, the most widely accepted of which are: **a)** Stochastic (ST) - which describes that wear and tear resulting from life span is the consequence of random molecular damage that alters the function and structure of cells, tissues and organs. The accumulation of these alterations would reach a level where some or all cells would be metabolically deficient to the point of death and **b)** Programmed (PT), based on the existence of a "roadmap" for sequential and coordinated events. Different patterns of metabolic control, mitochondrial damage, gene-cell signaling pathway interaction for apoptosis, telomere shortening and epigenetic events would also form a basis for changes throughout life (ARKING et al., 2000; MERCADO-SÂENZ et al., 2010).

Paredes Salido and Roca Fernández (2002) pointed out that, in addition to the genome, the production and release of energy, with the formation of reactive oxygen species (ROS) and environmental factors such as nutrition and physical exercise have been essential and determining elements in the process.

Based on these principles, and on the fact that *lifespan* is controlled, at least to a significant extent, by genetic pathways and biochemical processes that are evolutionarily conserved, hypotheses have been postulated and important biomarkers have been defined.

According to the Genetic hypothesis (GT), the intensity of biological aging is possibly regulated by genes that help maintain homeostasis and body function for longer than the average life expectancy and by those that affect various metabolic pathways, which can cause organic dysfunctions or increased morbidity and mortality among the elderly (JOHNSON, 2005). From this perspective, genetic factors are studied through the identification of *polymorphic* variants called SNPs (*single nucleotide polymorphism*) in a population, in which the variation of just one nucleotide in the DNA can have an impact on gene expression or regulation, interfering with the lifespan of the human species (FALLIN; MATTEINI, 2009).

The genetic contribution to aging and lifespan has been demonstrated using extensive twin registries and population-based studies. Heritability for twins has been estimated at 20.0 to 30.0% (HERSKIND et al., 1996; McGUE et al., 1993), while for populations, 15.0 to 25.0% (MITCHELL et al., 2001; KERBER et al., 2001). Studies also describe the relationship between genetic factors and lifestyle: the presence of moderate risks for cardiovascular disease in middle age and in the period between 60 and 69 years contributed to the survival of healthy and more frail individuals up to 85 years or more, respectively (DUTTA et al., 2011; TERRY et al., 2005).

Cruz (2014) proposed the existence of "survival" genotypes in response to environmental conditions such as dietary pattern, smoking and physical exercise, grouped into a series of genes with their by-products, in the post-reproductive period, which would provide plasticity to a given human life schedule, avoiding extinction. This could explain the inter- and intra-individual phenotypic differences between similar populations (KAZIMIROVA et al., 2004).

In lower organisms, the use of the *C. elegans* model broke through the challenges to establish the first assumptions about the genetic influence on lifespan. Marx (2002) described the discoveries of scientists about the genetic regulation of organ development and programmed cell death in this model organism and the existence of corresponding genes in humans.

Scientific evidence has also shown possible mechanisms involved in the shorter lifespan of progeroid syndromes, using murine and human models. These monogenic diseases are phenotypically characterized by early and rapid aging, with the *Hutchinson-Gilford and Werner* syndromes being classic examples. Genomic instability, caused by damage to DNA, followed by errors in its repair, would explain premature ageing (BURTNER; KENNEDY, 2010).

De Magalhaes and Church (2007) described that genes associated with ageing are conserved at the molecular level and there are multiple interactions with others: orthologous genes, associated with the life cycle, are found in model organisms and form a network of continuous connections and activity, and almost the majority of the proteins therefore expressed are implicated in age-related diseases (ARDs), which could explain the relationship between the appearance of diseases during this course.

The Oxidative Stress hypothesis, related to the Stochastic theory, states that the degenerative process, related to age, is a consequence of damage caused by reactive oxygen species, mainly from aerobic metabolism (SIMM; KLOTZ, 2015). The formation of reactive species, mainly oxygen, such as superoxide radical, hydrogen peroxide and hydroxyl radical, is caused by the catalytic action of enzymes during the transport of electrons in the respiratory chain and/or by exposure to different environmental factors (GOLUBEV, 1996). These molecules are electronically unstable and therefore extremely reactive, with the ability to combine with macromolecules such as lipids, proteins and DNA, causing cellular and molecular damage. This process is a type of pressure that induces a response from the body, mainly from antioxidant agents. However, during ageing, there is an exponential increase that exceeds the endogenous antioxidant capacity, in a variety of tissues and in different species, including humans (HARMAN, 1956).

The excess of reactive oxygen species leads to cellular oxidation, which is considered to be the initial stage of the diseases that accompany senility. ROS alter the molecular characteristics of membranes, promote genetic mutations, alter homeostasis and contribute to the formation of secondary compounds such as short-chain aldehydes, biomarkers of this process (HARMAN, 1991; MATÉS et al., 1999).

Although research into aging has advanced, generating considerable knowledge about its causes and outcomes, it is still subject to countless debates and conflicting hypotheses. So far, even with the gains in life expectancy seen in recent decades, the range of diseases affecting the elderly has not changed. Even so, many can reach extreme ages, longevity, with the presence of these dysfunctions or, on the other hand, without their presence, less frail and with satisfactory cognitive and physical functions (CHRISTENSEN; JOHNSON; VAUPEL, 2006).

Newman and Murabito (2013) defined healthy ageing as survival to specific age groups, greater than or equal to 85 years. However, slowing down the ageing process is a daunting task because it involves the accumulation and integration of countless pieces of information from a biological, physiological and psychosocial point of view.

Scientific research that aims to find new ways of caring for the health of the elderly, especially the long-lived, and that seeks to integrate the different types of information about the ageing process, is extremely important in order to alleviate the incapacity and physical and organic frailties that accompany the majority of these individuals.

Chapter 3 Human longevity: concept, phenotypes and determining factors

The dramatic increase in the elderly population, especially the "oldest old", has aroused the interest of scholars in aging and the advancement of research in biogerontology. It is becoming increasingly important not only to study multimorbidity, but also the phenotypes and possible biological mechanisms of healthy ageing (DAVINELLI; WILCOXX; SCAPAGNINI, 2012).

Longevity is a complex phenomenon and results from the interaction of numerous genetic, environmental and behavioral factors. In studies of the genetics of human ageing, the term "longevity" refers to survival that exceeds the estimated maximum life span of a population. From a chronological point of view, this would mean reaching 85 years of age or more, with this age group being referred to as the very old or *oldest-old.* Considering physiopathological aspects, longevity phenotypes are characterized by the combination of older age, the absence of multiple diseases and acceptable physical and cognitive capacities, which imply the need to add another name: "healthy ageing" or "healthy longevity" (BROOKS-WILSON, 2013).

According to Mercado-Sàenz et al. (2010), lifespan and healthy lifespan are closely related. However, human beings have been successful in terms of their lifespan, but have made little progress in terms of their *healthspan*. In the last two decades, there has been a greater likelihood of reaching longevity, but with negative health and well-being phenotypes or a relatively short senescence phase, free of chronic degenerative diseases and desirable physical and functional performance, which is related to "healthy ageing".

In an attempt to understand these different phenotypes, genomic geriatrics has been investigating possible genes and metabolic pathways associated with ageing and candidates for longevity. The first findings, from large family or population-based studies, showed that genetic inheritance is a factor that contributes to defining life span.

Studies with twins, mono-twins and dizygotic twins have shown that approximately

25% of the variation in survival is determined by genetic characteristics (HERSKIND et al., 1996; HJELMBORG et al., 2006). Genetic influence was also significant when assessed among siblings of long-lived people, as they were around four times more likely to become octa or nonagenarians than siblings of *young-old* people (CHRISTENSEN; JOHNSON; VAUPEL, 2006; PERLS et al., 1998). The genetic effects on the lifespan of twins were minimal up to the age of 60, but became significant and stable with age, being accentuated from the age of 85 (HJELMBORG et al., 2006).

Similarly, in population-based studies, peculiar characteristics were observed: as age increased, the risk rate of developing diseases such as cancer, diabetes, cardiovascular, respiratory and neurodegenerative diseases was substantially lower, in addition to greater genetic participation at more extreme ages (BROOKS-WILSON, 2013; FORTNEY et al., 2015; MITCHELL et al., 2001).

The genetic contribution to human aging and longevity is possibly due to the interaction of several genes, each with its own particular, connected and pleiotropic effects. There are *longevity predisposition genes*, genetic variants or gene regulation, which help maintain homeostasis and body function for longer than the expected average, exceeding life expectancy, while another group, *frailty genes*, would increase susceptibility to age-related diseases, with early morbidity and mortality, by affecting metabolic pathways, culminating in potential organic dysfunctions (MURABITO; YUAN; LUNETTA, 2012).

Genetic association studies have achieved important results in the identification of single nucleotide variations (SNPs) in the DNA sequence, related to phenotypes characteristic of the aging process and show potential discoveries for new metabolic and signaling pathways (BROOKS-WILSON, 2013; FIGARSKA; VONK; BOESEN, 2013; JACOBSEN et al., 2010; SEBASTIANI et al., 2013; WILLCOX et al., 2008).

Considering longevity as a polygenic trait and its polymorphic variants, Barzilai and Gabrielly (2010) proposed possible changes in the genotype profile over the life course of elderly individuals, from 60 to 100 years of age, assuming that this is the maximum life span among populations (Figure 2). SNPs that decline as age increases are related to *frailty genes* and an increase in the proportion of morbidity and mortality, while a different and more relevant pattern would be observed when SNPs increase with age, which would be related to *longevity predisposition genes,* protective against ageing diseases. The results of Sebastiani and Perls (2012) corroborate this, as they suggested that in centenarian women and men, the genetic contribution would reach 33% and 48%, respectively. However, there are genotype profiles which do not change in frequency over the course of life and which are not related to longevity.

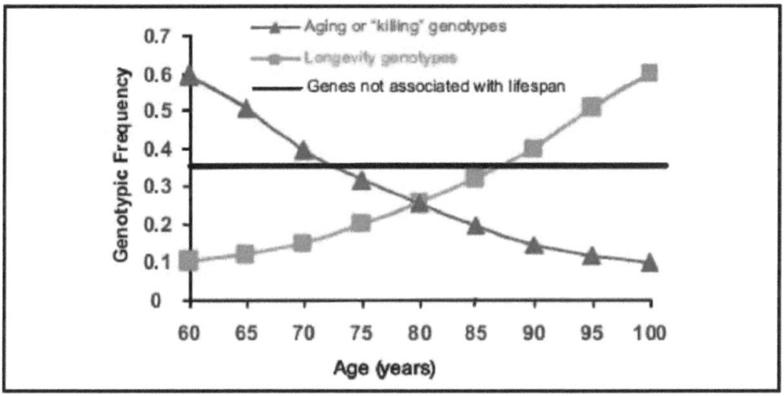

Figure 2: Changes in genotype frequencies as a function of age. Knowing that the "x" axis refers to lifespan in years and the "y" axis to genotype frequency, despite reaching the lifespan threshold, there is significant selection. Green lines represent Iongevity predisposition genes - genotypes that contribute to Iongevity and protection against diseases that accompany ageing. Red lines represent frailty genes - genotypes associated with increased disease prevalence and mortality. Black lines indicate genes with SNPs that do not change their frequency with age (BARZILAI; GABRIELLY, 2010).

These findings were driven by the discovery of evolutionarily conserved genetic variants related to longevity in invertebrates, which are short-lived and adept at controlling genetic and environmental factors. Variations in the gene called *age-1*, a gerontogene found in the nematode *C. elegans,* are thought to be responsible for increasing lifespan by two to five times and by up to eight times when combined with calorie restriction (HOUTHOOFD et al., 2004; JOHNSON, 2005).

Other animal models have also contributed to biogerontology, such as those developed with *Drosophila melanogaster,* which identified the *Tor* gene, which codes for a protein modulated by the availability of amino acids and, therefore, by diet.

Furthermore, it is important to mention that animal studies have revealed that longevity genes are age-specific: variants in the *age-1, spe-10 and clk-1* genes appear to be involved in reducing mortality in *C. elegans* only in late life or low reproductive stages (JOHNSON et al., 2001; KAPAHI; ZID, 2004).

Mice *knocked out* for growth hormone (GH) receptors and for the insulin signaling pathway IIS (*insulin/Igfl-like signaling)* can substantially increase their lifespan. This gene ablation increases the expression of antioxidant enzymes and the stress resistance of myocytes and fibroblasts, activates transcription factors called FOXO, and reduces the onset of ageing diseases (BARTKE, 2005).

The successful development of this research has significantly facilitated efforts to identify genetic variations in human pathways.

In addition to the genetic component, it is known that the heterogeneity of longevity phenotypes is due to environmental and behavioral factors. Scientific evidence shows that education, income, ethnicity, physical exercise, smoking, alcohol consumption and social interaction are all factors that have been linked to healthy ageing (HAVEMAN-NIES, de GROOT, STRAVEREN, 2003; SWINDELL et al., 2010; WILLCOX, HE, CHEN; 2006).

Colman et al. (2014) and Barzilai and Bartke (2009) have shown an association between calorie restriction (CR), a 25 to 50% reduction in calorie intake, without malnutrition and maintaining nutrients essential for the proper functioning of the body, with increased life span and a decline in the diseases inherent in the aging process. It is known that, in mammals, the protein Sirtuin 1 (SIRTl) is involved in the metabolic response under CR conditions and polymorphic variants of its gene (*SIRTl*) are involved in longevity. In addition, according to Akbaraly et al. (2013), Hodge et al. (2014) and Assmann et al. (2015) the intake of fruit and meat rich in SIRTl is associated with longevity.

Iipids correlate positively and negatively, respectively, with healthy aging.

Clinical research shows that lifestyle changes, such as calorie restriction, can slow down intrinsic aging and lead to favorable outcomes in terms of variables such as body mass index, lipid profile, glycemia and insulin sensitivity markers (FONTANA; KLEIN, 2007).

Knowledge of these possible predictors of survival and health is very important for contemporary research into longevity, as they can be modified, providing "very old" individuals with a better quality of life and a lower incidence of chronic degenerative diseases. Elucidating how genetic factors and their interaction with environmental components contribute to longevity and ageing phenotypes is one of the fields explored by genomic geriatrics and which has aroused the interest of researchers (MURABITO; YUAN; LUNETTA, 2012). It is possible that a healthy lifestyle leads to successful ageing and/or *longevity predisposition genes* protect older people from the harmful effects of inadequate habits (SWAPNIL et al., 2011).

Similarly, gender has influenced the length of survival. In general, women have a longer average lifespan than men, accounting for the largest proportion of the world's "elderly". Some hypotheses for this difference are hormonal characteristics, the compensatory factor of the X chromosome in women, the higher occurrence of diseases such as cardiovascular diseases and accidents in men (NEWMAN; MURABITO, 2013). On the other hand, the effects of gender on longevity can be modified by environmental factors as well as genetic ones (BROOKS-WILSON, 2013; CATALANO, BRUCKNER, SMITH, 2008). Poulain, Peres and Salaris (2011) showed that in a population of Iongevos on the island of Sardinia, factors such as diet, physical exercise, family and social life, as well as genetic/biological characteristics, contributed to a lower mortality rate among men, contrary to expectations.

The phenotypic variability observed among elderly individuals and longevity can also be determined by epigenetic characteristics, which have been investigated more recently in experimental models, mammalian cell culture and human studies (DEELEN et al., 2014; MCCAULEY; DANG, 2014). Epigenetic changes involve chemical alterations in DNA molecules or histones, as well as the production of a special class of short, non-coding ribonucleic acid (RNA), the microRNAs. These posttranscriptional mechanisms are biologically important because they play a role in regulating gene expression and are considered relevant candidates for elucidating the effects of environmental factors on ageing (SIMMONS, 2008).

In Brazil, there are few studies on genomic geriatrics and longevity. Most studies have focused on studying socio-demographic conditions, clinical/biochemical/anthropometric characteristics that accompany DRI, functional capacity and the profile of caregivers (VALADARES; VIANNA; MORAES, 2013; LIMA; MENEZES, 2011). Research into polymorphic variants of longevity predisposition genes has been or is being carried out mainly in states in the South and Southeast regions (Table 1). These data show the growing need for research in the field of biogerontology, which deals with the possible genetic and environmental factors involved in increasing the lifespan of our population.

Table 1. Studies carried out in Brazil on Human Longevity and Genetic Variants.

Study	*State*	*n*	*Gene (*SNP*)*	*References*
GENESIS	RS	489	*SOD2* (rs4880)	Taufer et al. (2005) do Valle Gottlieb et al. (2011)
VERANOPOLIS	RS	64 252	*APOE* (rs429358/rs7412) *SOD2* (rs4880)	Schwanke et al. (2002) Da Cruz et al. (2004)
BAMBUI	MG	1406	*APOE* (rs429358/rs7412)	Lima e Costa et al. (2000) Fuzikawa et al. (2008)
				Lebrao and Laurenti (2005)
KNOW	SP	604	*APOE* (Exome/NGS)	Naslavsky (2015)-personal communication

n = sample number; *SNP = single nucleotide polymorphism* ; *NGS = Next Generation Sequencing.*

Many metabolic pathways, such as responses to the hormone insulin, oxidative stress, lipid metabolism and calorie restriction/energy expenditure, appear to be crucial for achieving longevity (CHRISTENSEN; JOHNSON; VAUPEL, 2006). It is known that there are genes and polymorphic variants related to these signaling pathways, which have been considered natural candidates and have a notable impact on lifespan, which corroborates the hypothesis that there are genes predisposing to longevity and frailty.

The elucidation of the genetic basis, as well as oxidative and genomic damage in aging and its role in physical, biochemical and pathological changes over the course of life remains intriguing and complex, offering gerontology a broad spectrum of research with a view to obtaining possible predictors that add quality of life to these years added to humanity.

Chapter 4 Biomarkers of Human Longevity

Considering the scenario of an ageing population and an increase in life expectancy, combined with the human motivation to preserve life to the fullest and with quality, as well as the substantial costs to health systems of chronic degenerative diseases and the multiple factors involved in this process, it is of the utmost importance to study possible predictors or markers that interfere with longevity.

Biomarkers are defined as characteristics that can be measured and evaluated, being indicators of normal or pathogenic biological processes, or even of pharmacological response to therapeutic intervention (DE ZWART et al., 1999; LABAER, 2005).

For the study of ageing and longevity, various biomarkers have been proposed, such as genes and genomic regions identified in *GWAS (Genome-Wide Association Studies)* studies (BROER et al., 2015; FORTNEY et al., 2015), as well as clinical, biochemical and anthropometric parameters and environmental factors (DUTTA et al., 2011).

In addition, studies have shown that as we age, cellular dysfunction and senescence occur in our bodies, partly due to an exacerbated increase in the production of reactive oxygen species, which can react with macromolecules and also culminate in the production of secondary cytotoxic or genotoxic compounds, such as short-chain aldehydes (BECKMAN; AMES, 1998; DEL RIO; STEWART; PELLEGRINI, 2005). These carbonyl compounds, unlike ROS, are relatively stable and diffuse into the extracellular environment, reaching targets far from their site of origin, propagating oxidative damage.

Considering these factors and events as critical determinants of lifespan, estimating them in populations that have not yet been studied can improve scientific knowledge and help develop strategies related to biogerontology that seek to improve quality of life in order to achieve longevity. More effective preventive or diagnostic actions could be taken according to the profile of each elderly individual, which would have a significant impact in terms of public health.

Chapter 5 Genes associated with Human Longevity

As previously described, the genetic component contributes approximately 25% to lifespan and genes predisposing to longevity have been identified in studies with animal models and humans. Genes such as *FOXO3, SOD2, APOE and SIRT1* have been described as modulators of longevity and are involved in Iipid and carbohydrate metabolism pathways, the response to oxidative stress and susceptibility to the development of DRIs (BARZILAI et al., 2012; DATO et al., 2013).

FOXO3

The *FOXO3* gene *(Forkhead box O3),* located on chromosome 6q21, encodes the multifunctional transcription factor *forkhead box O3*, which belongs to the family of evolutionarily conserved FOXO, FOXO1, FOXO4, FOXO6 proteins, from *Caenorhabditis elegans* to mammals (LEE et al., 2003). FOXO3 is involved in processes such as cell proliferation, apoptosis, response to oxidative stress, cell cycle regulation, metabolism and longevity (TZIVION; DOBSON; RAMAKRISHNAN, 2011). This transcription factor is expressed in tissues that respond to the hormone insulin, adipocytes, skeletal muscle, liver, pancreas and has been considered a key mediator in the insulin signaling pathway *IIS (Insulin/Igf1-like signaling)* (LAM; FRANCIS; PETKOVIC, 2006).

Although the precise mechanisms have yet to be elucidated, *in vitro* studies have shown that the *IIS* pathway can affect lifespan by various mechanisms.

When the *IIS* pathway *is* inhibited by the FOXO3 factor, the transcription of heat shock protein genes involved in the stress response is activated (MURPHY et al., 2003). FOXO3 can also act to arrest the cell cycle and stimulate DNA repair mechanisms, which reduces cell death and the risk of malignancy, limiting the ageing process (TRAN et al., 2002). There is also evidence that FOXO factors are involved in the expression of antioxidant enzymes. FOXO3 appears to act as a transcription factor for multiple stress resistance pathways, in response to decreased signaling from the *IIS* pathway.

Di Bona et al. (2014) defined the *Insulin/Igf1-like* pathway as a cascade of phosphorylations that regulate nuclear translocation and the activation of FOXO proteins, including FOXO3. In response to exogenous factors (diet) or growth hormone, the pancreas secretes insulin and the liver produces insulin-like growth factor 1 (IGF-1). The binding of IGF-1 to specific receptors inhibits the FOXO3 protein, which prevents the expression of many homeostatic genes, such as those of the antioxidant and immune systems, involved in longevity (Figure 3). However, the expression of the *FOXO3* gene can be activated by proteins called sirtuins, NAD-dependent deacetylases$^+$ (nicotinamide adenine dinucleotide), considered anti-aging molecules. Specific polymorphic variants may also be involved in modulating this pathway, such as the SNP rs2802292 G/T, located in the intronic region of the gene.

In FOXO3 *knockout* mice, insulin resistance, hematological abnormalities, inflammatory processes and infertility have been observed (CASTRILLON et al., 2003; KERDILES et al., 2010).

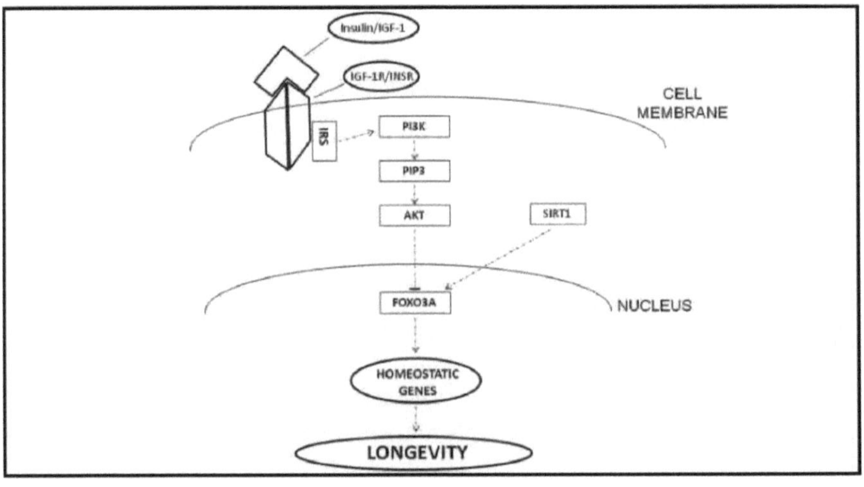

Figure 3: *Insulin/Igf1-like* signaling pathway *and* its implications for longevity. This signaling pathway plays a critical role in determining longevity. The binding of IGF-1 to specific receptors in a cascade of phosphorylations inhibits the FOXO3 protein, which prevents the expression of many homeostatic genes involved in longevity. The SIRT1 (Silent Information Regulation Type 1) protein, on the other hand, activates this transcription factor. Genetic variants that reduce this signaling have been identified: some studies have shown this association with a longer lifespan (DI BONA et al., 2014).

Willcox et al. (2008) were the first researchers to identify that the *FOXO3* gene was strongly associated with longevity in a Hawaiian population of long-lived men of Japanese origin. The rs2802292 genetic variation showed that these long-lived men had one or more copies of the G allele and that controls had a higher frequency of the T allele. These elderly men, compared to controls, also had phenotypes related to healthy ageing, such as a lower prevalence of cancer and cardiovascular disease, a better lipid profile and lower glycaemic and insulin levels.

Subsequent studies, with sample sizes ranging from 600 to 1800 and an age range of 18 to 110 years, confirmed these results: Anselmi et al. (2009) and Soerensen et al. (2010), in Caucasian populations, also validated this association when they analyzed data from men and women separately. Similarly, a study by the latter research group with a Danish population, which considered the characteristics cognitive function, functional capacity and self-reported health, showed a positive association between *FOXO3* SNPs, including rs2802292, and longevity (SOERENSEN et al., 2015).

SOD2

Superoxide dismutases (SODs) are a class of enzymes that catalyze the conversion of superoxide anions ($O_2^{\cdot -}$), a reactive oxygen species, into hydrogen peroxide (H_2O_2), which is subsequently converted into water (HALLIWELL, 1978). These ROS are generated mainly during the transport of electrons in oxidative phosphorylation. SODs play an important role in biological oxidation-reduction pathways and are crucial in the endogenous antioxidant defence of the signalling system that protects cells and tissues from oxidative stress, characterized by an increase in the production of reactive compounds (NOZIK-GRAYCK; SULIMAN; PIANTADOSI, 2005).

In mammals, there are three identified isoforms of SOD, with different cellular distribution: SODI or CuZn-SOD (cytoplasmic - copper and zinc dependent), SOD2 or Mn-SOD (mitochondrial and manganese dependent) and SOD3 or EC-SOD (extracellular) (ZELKO; MARIANI; FOLZ, 2002). Superoxide dismutase 2 is responsible for our body's primary antioxidant defense and is encoded by the *SOD2* gene, located on chromosome 6q25.3. Its role in aging, accompanying diseases and longevity has been described, with the rs4880 C/T SNP being the most extensively studied (BUDOVSKY et al., 2013). Holley et al. (2012) and Kaminski et al. (2013) also emphasized that this enzyme is responsible for maintaining moderate levels of superoxide and hydrogen peroxide in the intracellular environment, which are positive for homeostasis.

The rs4880 polymorphic variant, with a substitution of the amino acid valine for alanine, causes a conformational change in the protein's structure from β-pleated to α-helix, which interferes with its ability to neutralize O_{2^-} radicals$^-$. SOD2 α-helix (Cytosine) would be related to a 30 to 40% increase in its activity, while the β-pleated conformation (Thymine), with a reduction in its enzymatic efficiency (SHIMODA-MATSUBAYASHI et al., 1996). On the other hand, Sutton et al. (2005) suggested that the metabolic imbalance in SOD2 activity, dependent on these variants, was more related to alterations in its transport within the mitochondria. The CC genotype gives the enzyme the ability to cross the inner and outer mitochondrial membranes rapidly towards the matrix, whereas TT would cause a stationary phase of SOD2 between the membranes.

The contribution of this SNP in determining the processes of ageing, disease and longevity corroborates the results of the functional studies mentioned above. However, the findings have been conflicting. Soerensen et al. (2009) evaluated the association of the rs4880 SNP with longevity in a Danish population of 1650 individuals and observed that those carrying the C allele showed a decrease in mortality. In Brazil, a study was conducted in Porto Alegre, RS, with 489 individuals (newborns, adults and the long-lived). No association was found between genetic variants and longevity. However, the CC genotype was related to an increased risk of prostate and breast cancer and genomic damage in peripheral blood cells (TAUFER et al., 2005). The studies by Stessman et al. (2005) and Tian et al. (2011) showed an increase in the frequency of T alleles with age and with the prevalence of *diabetes mellitus.* In the study by Gentschew et al. (2013), no association was found for rs4880 and lifespan between 1612 long-lived (> 95 years) and 1104 Caucasian controls (60-75 years).

The use of lower and murine organisms also makes the relationship between *SOD2* and lifespan evident. Overexpression of the *SOD2* gene in *Drosophila melanogaster* resulted in a proportional increase in its lifespan and enzymatic activity (SUN et al., 2002). In a study in which only 50% of SOD2 was expressed, oxidative damage and impaired mitochondrial function were observed (HINERFELD et al., 2004).

Other environmental factors may be involved in these different phenotypes, since this SNP generated results that are impacted by conditions such as antioxidant diet and physical exercise. The intake of fruit and vegetables, a source of antioxidants, reduced the effects of the risk of developing breast cancer in premenopausal women carrying the CC genotype (risk factor) (AMBROSONE et al., 1999). Bresciani et al. (2013) showed that oxidative stress from physical exercise is modulated by the rs4880 SNP. The relationship between the polymorphic variants and measures of protein content, enzyme activity and gene expression in 65 adult males was evaluated and it was observed that, one hour after exercise, for the CC genotype, Mn-SOD content was higher, as well as *SOD2* expression levels.

APOE

Apolipoprotein E (ApoE) is a 34KDa glycoprotein that plays an important role in lipid homeostasis in the brain and peripheral tissues (MAHLEY, 1988). This protein participates in the transport and redistribution of cholesterol and is predominantly synthesized by the liver. The gene that codes for it, *APOE,* is polymorphic and is located on chromosome 19q13.2 (DAS et al., 1985). The combination of two single nucleotide polymorphisms, rs429358 and rs7412, in exon 4, and their respective differences in the 112 and 158 amino acid residues, define three isoforms for *APOE*: ε2, with cysteine residues in both sites (cis112, cis158); ε3, with cisterna and arginine (cis112, arg158) and ε4, with arginine residues in both positions (arg112, arg158). Figure 4 shows this schematically. Considering these isoforms, six possible diplotypes are observed, ε2/ε2, ε2/ε3, ε2/ε4, ε3/ε3, ε3/ε4 and ε4/ε4 (ZHONG et al., 2016).

These variants and, consequently, the differences between the amino acids cysteine and arginine alter the structure of ApoE and its function: ApoE2 and ApoE3 preferentially bind to high-density lipoprotein (HDL-c), while ApoE4 adheres to low-density lipoprotein (LDL-c) and very low-density lipoprotein (VLDL-c) (SAITO et al., 2003).

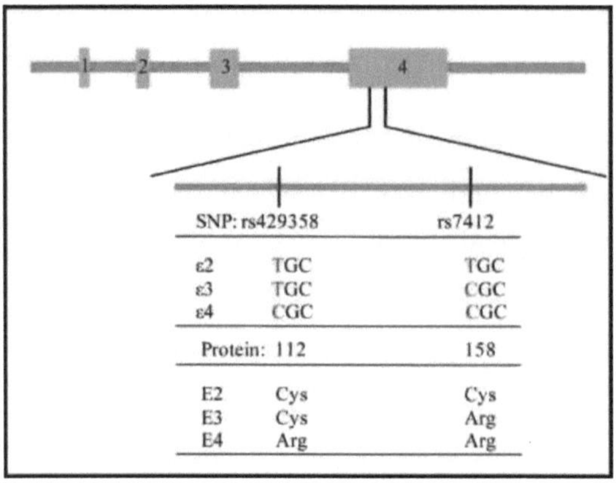

Figure 4: Schematic representation of the location of SNPs rs429358 and rs7412 in exon 4 of the *APOE* gene and their possible allele and protein isoforms, according to changes in the residues of amino acids 112 and 158 (Adapted from Zhong et al., 2016).

The presence of the cisterna residue in ApoE2, at position 158, causes a rearrangement in the existing bond between arg154-arg158, replacing arg158 with arg 150, which changes the protein's binding potential to lipoprotein receptors (MAHLEY; HUANG; RALL, 1999). On the other hand, in ApoE4, the substitution of cisterna for aginine at position 112 leads to an interaction between arginine and glutamine, at positions 61 and 255, respectively, in aqueous media, culminating in receptor-binding domains in the ApoE protein that do not exist in the E2 and E3 isoforms (ZHONG; WEISGRABER, 2009). In turn, these changes make the structure of E4 more compact and with less stable and protected regions compared to E3 (HUANG et al., 2011). As a result, ApoE2 and E4 have receptor binding profiles that favor hypocholesterolemia and hyperlipidemia, respectively (MAHLEY; RALL JR, 2000).

These biochemical and metabolic outcomes are responsible for the association of these isoforms with different biological events such as oxidative stress and the inflammatory process (HUEBBE et al., 2011).

According to this conjecture, human studies on *APOE* genetic variants, lifespan and diseases affecting the elderly have been widely carried out. The ε4 allele has been associated with early mortality, cardiovascular and neurodegenerative diseases and ε2 with longevity and healthy life span (CHRISTENSEN; JOHNSON; VAUPEL, 2006; DEELEN et al., 2011; SCHUPF et al., 2012). The population distribution of ε3, ε4 and ε2 around the world is around 77.9%, 13.7% and 8.4%, respectively (CHRISTENSEN; JOHNSON; VAUPEL, 2006).

Soerensen et al. (2013), in a study involving Caucasians, with 1089 nonagenarians and 736 middle-aged controls, showed a positive association between the *APOE* gene and life span extension. However, Tindale et al. (2014) found a higher frequency of the ε4 allele among controls, but not of ε2 for the very elderly, in a study of 752 healthy controls and Iongevins. In a population of 1,406 elderly people aged 60 or over in the southwest of Minas Gerais, it was observed that there was no association between hypertension and *APOE*, but individuals carrying the ε4 allele had high levels of LDL-c, while the ε2 allele was related to lower levels of this lipoprotein (FUZIKAWA et al., 2008).

SIRT1

The *SIRT1* gene (*Silent Information Regulator Type 1*) located on chromosome 10q21.3 consists of 11 exons and encodes a protein with an important role in longevity and health promotion, sirtuin 1 (SIRT1). The sirtuins identified in humans constitute a family of seven enzymes, which are evolutionarily united by the presence of a catalytic domain, binding to nicotinamide adenine dinucleotide, verified via deacetylase-NAD activity$^+$, primarily in the *SIR2* gene (*Silent Information Regulator* 2) of *Saccharomyces cerevisiae*, involved in extending the life span of this organism (FRYE, 2000; SMITH et al., 2000).

SIRT1 is the most studied member of this family because it is the closest homologue to Sir2 and because of its role in regulating metabolic processes and the cell cycle, in the response to oxidative stress and in ageing (HALL et al., 2013; MICHAN; SINCLAIR, 2007). This enzyme is found in the nuclear compartment of liver, pancreatic, muscle, heart, adipocyte and neuron cells. It is activated by high levels of the NAD^+ / NADH ratio, a key indicator of oxygen consumption and, consequently, low energy availability (NOGUEIRAS et al., 2012). By catalyzing reactions involved in the removal of acetyl groups from lysine residues of proteins such as transcription factors, which induce catabolic processes, this enzyme coordinates the increase in energy stocks in cells, ensuring homeostasis (Figure 4).

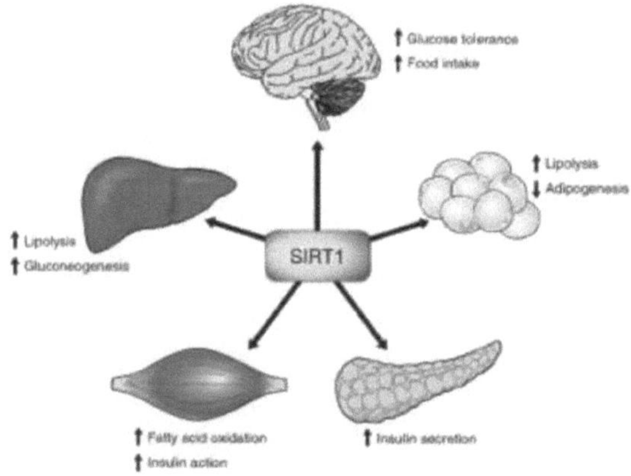

Figure 5: Metabolic roles of SIRT1 in peripheral tissues and the central nervous system (NOGUEIRAS el al., 2012).

Research in animal models has shown that elderly transgenic mice with moderate expression of SIRT1 had a better general state of health, reduced expression of genes related to ageing phenotypes and a lower incidence of spontaneous carcinomas and sarcomas (HERRANZ et al., 2010). Similarly, strains of *Sirt-Itg* mice, in a model of heart disease, showed a reduction in the levels of markers of ageing and cardiac hypertrophy, as well as satisfactory resistance to stress (ALCENDOR et al., 2007).

Considering that SIRT1 is a guardian of the healthy lifespan of mammals in diverse disease contexts, genetic variants of *SIRT1* such as the rs2273773 C/T SNP have been investigated in the context of longevity, being an area of great interest and questioning.

Figarska, Von and Boezen (2013) examined the association between the frequency of SNP rs2273773 genotypes and lifespan in 1390 Danish individuals (49.4 - 61.8 years), but found no positive results. Prospective studies by Kuningas et al. (2007) and

Flachsbart et al. (2006) also found no association between this variant and longevity in a sample of around 1000 very elderly Caucasians. However, the first group of researchers observed that *SIRT1* may be related to satisfactory cognitive function. Similar findings were described by Soerensen et al. (2013), who studied candidate genes for longevity in a population of Caucasians.

There are few studies on the study of *SIRT1* SNPs and lifespan in humans. In recent years, findings have largely shown a relationship between these variants and phenotypes associated with metabolism and morbidities of ageing (LAM et al., 2013; NOGUEIRAS et al., 2012). Han et al. (2015) showed that genetic variations in *SIRT1* are associated with the lipid profile of the elderly, but not with longevity. A relationship between polymorphic variants of this gene, increased risk of developing diabetes, reduced insulin sensitivity and Body Mass Index was observed in a population of Indians (DONG et al., 2011). In another study, elderly carriers of the CT genotype (rs2273773) had higher levels of SIRT1, indicating relative protection to the C allele. However, at the same time, a higher level of oxidative stress was observed in this group (KILIC et al., 2015).

However, the above does not exclude the possibility that polymorphic variants of *SIRT1* modulators or substrates of its protein may play a critical role in extending lifespan in humans. To test this hypothesis, experimental models that are able to maintain NAD^+ concentrations at high levels, such as those inducing caloric restriction, have been conducted and have shown the consequences of increasing SIRT1 on DRI-related phenotypes in different tissues, with increased lifespan and better Sir2 activity in lower organisms, as well as positive effects on human health (LIN et al., 2000; RIBARIC, 2012).

Chapter 6 Oxidative Stress, Malondialdehyde and the Human Lifespan

The inverse relationship between increased production of reactive oxygen species and maximum life span has been demonstrated in several studies, and this event is one of the main causes of ageing and the diseases that accompany it (DATO et al., 2013; SOHAL; WEINDRUCH, 1996).

Since the 1950s, the Reactive Oxygen Species Theory of Ageing has postulated that this process is partly due to cellular and molecular oxidative damage caused by cumulative exposure to ROS (HARMAN, 1956). As previously described, oxidative stress is the result of an imbalance between pro-oxidant and antioxidant states and ROS, their products, are substantially unstable and can react with DNA, proteins and lipids.

Of these biological targets, lipids and the process of lipid peroxidation have been the most extensively studied. The abundant presence of phospholipids in cell membranes, and in sites of reactive oxygen species production, make them easily accessible and rapidly affected endogenous targets (DE ZWART et al., 1999). In particular, polyunsaturated fatty acids (PUFAs) are extremely susceptible to reactions with ROS, generating lipid peroxides. This process, called lipid peroxidation, can alter the fluidity and plasticity of the membrane of various cells such as neurons, heart and muscle cells, and cause damage to their integrity (PRATICÓ, 2000). In general, the sensitivity of fatty acids to oxidation increases exponentially depending on the number of double bonds in their molecule, which are prone to attack (BIELSKI; ARUDI; SUTHERLAND, 1983).

In this autooxidative process, ROS, mainly the hydroxyl radical ($OH^{\cdot}$), react with polyunsaturated fatty acids (LH), forming fatty acid free radicals ($L^{\cdot}$), which react spontaneously with oxygen, producing fatty acid peroxireadicals ($LOO^{\cdot}$). The latter can abstract a hydrogen atom from another fatty acid molecule, forming lipid hydroperoxides (LOOH) and a new radical, which sets off a chain reaction. Hydroperoxides can undergo decomposition reactions and be converted into secondary products of low molecular weight, such as 4-hydroxy-2-nonenal, acrolein and propanedial or malondialdehyde - MDA (Figure 3) (AUGUSTO, 2006; DEL RIO; STEWART; PELLEGRINI, 2005).

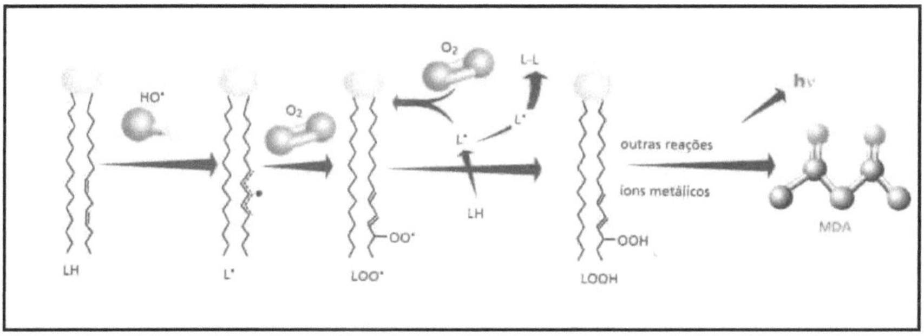

Figure 6 - Stages of Lipid Peroxidation (Modified from Augusto, 2006).

These bioactive compounds have received particular attention because they can propagate and amplify oxidative injury, since they are relatively stable, have a remarkable half-life and diffuse easily through the membrane, reaching sites other than those of origin (NEGRE-SALVAYRE et al., 2008). In addition, they are detected in biological tissues and have the ability to react with nucleic acids and proteins, altering mechanisms involved in cell functionality (DEL RIO; STEWART; PELLEGRINI, 2005). *In vitro* and *in vivo* research has demonstrated the presence of these secondary products in various models of pathological conditions that accompany ageing (MOREAU et al., 2005; PAMPLONA et al., 2005; UCHIDA, 2000).

Malondialdehyde is the most widely used biomarker for monitoring lipid oxidation, as well as aging *per se* or healthy aging. This carbonyl compound is derived from the endocyclization of linoleic, linolenic and docosahexaenoic fatty acids. It can be found in its free form or bound to the sufhydryl or amino groups of molecules and its measurement in plasma is a reflection of the systemic state (ESTERBAUER; SCHAUR; ZOLLER, 1991; MUTLU-TURKOGLU et al., 2003).

Various methodologies have been developed to quantify MDA in biological samples. The most commonly used method is the TBARS test (thiobarbituric acid reactive substances), introduced by Yagi et al. (1976), which allows a pink chromogen to be read by spectrophotocolorimetry in plasma samples. However, this test is unspecified because it takes into account interferents in the measurement (STEGHENS et al., 2001).

In contrast to this limitation, the High Performance Liquid Chromatography (HPLC) used to identify and quantify malondialdehyde is more specific, reproducible and reliable. Derivatizing agents such as 2,4-dinitrophenylhydrazine (DNPH) are often used to make the identification of MDA more precise and diode array spectrophotometric (DAD) detectors increase selectivity. With this in mind, the study by Antunes et al. (2008) developed and validated a methodology for quantifying plasma MDA by HPLC in Brazilian individuals aged between 20 and 40, using alkaline hydrolysis, acid precipitation, DNPH and DAD. The levels of this carbonyl compound ranged from 2.93 to 3.69 μmol/l of plasma.

Nielsen et al. (1997) used this test and similar criteria to analyze MDA in human plasma from Danish individuals aged 20 to 79 and found values of 1.19 to 1.76 μmol/l of MDA in plasma for elderly individuals aged 69 to 79. Belenguer Varea et al. (2015), using CLAE as a procedure, observed in a sample of 59 individuals that nonagenarians and centenarians had lower levels of MDA (1.44 ± 0.45 μmol/l) compared to controls aged 70 to 80 (1.84 ± 0.59 μmol/l). Studies have shown that long-lived individuals may have protective genetic signals involved in a better antioxidant response (LABAT-ROBERT; ROBERT, 2014; MECOCCI et al., 2000).

Chapter 7 Genomic Integrity, Aging and Longevity

The DNA molecule plays a critical role in maintaining the homeostasis and survival of living beings, as it is involved in the synthesis of cellular components, the cell cycle and its signaling pathways, apoptosis, genomic stability and senescence (LOMBARD et al., 2005). Thus, the study of possible cumulative and age-related damage to its molecule deserves special attention and has shown its association with ageing and increased morbidity and mortality (SOARES et al., 2014).

It is known that, in addition to direct oxidative damage to macromolecules, the dramatic increase in ROS, accompanied by a deficient antioxidant system, can generate secondary genotoxic products such as malondialdehyde, as well as influencing DNA repair mechanisms (WILSON; BOHR; MCKINNON, 2008).

Reactive oxygen species can cause DNA-protein cross-linking, damage to the phosphate-deoxyribose structure, as well as specific chemical modifications to purine and pyrimidine bases, possibly resulting in mutations, broken nitrogenous base bonds or DNA strands (DE ZWART et al., 1999; DIZDAROGLU, 1991). In addition, the carbonyl compound MDA, a product of lipid peroxidation, can form adducts or covalent bonds with the nitrogenous bases of deoxyribonucleic acid (MARNETT, 1999).

The levels and extent of oxidative damage to DNA in biological fluids such as plasma and tissues have been considered reliable biomarkers. Among the various proposals, the Comet Assay is one of the simplest and most widespread techniques for genotoxicological studies in eukaryotes (NERI et al., 2015).

The Comet Assay or SCGE Assay ("*Single Cell Gel Electrophoresis*"), an electrophoresis of individual cells, was first described by Ostling and Johanson (1984), is fast, economical, sensitive, requires few cells for analysis and evaluates genomic damage which, after being processed, can culminate in mutations. It combines biochemical techniques to detect single strand breaks, alkali labile sites, abasic sites, excision of incomplete repair sites, double strand breaks and cross-links, with the typical cell analysis approach of cytogenetics (COLLINS, 2004; SINGH et al., 1989;

TICE et al., 2000).

In this test, a cell suspension is mixed with agarose and lysed using a detergent and high salt concentrations. The cell membranes are lysed, histones are extracted and the genetic material, which is larger, heavier and adhered to the residual nuclear matrix, fills the space of the gel in a nucleus-like structure called a nucleoid. When subjected to the electrophoretic field, the DNA migrates, depending on the extent of the damage, towards the anode, with the appearance of a comet, which can be visualized under a fluorescence microscope and classified according to the fragments, length and intensity of the tail (Figure 7) (COLLINS, 2004; FAIRBAIRN; OLIVE; O'NEILL, 1995).

Figure 7. Comet classes that can be found in the analysis: (a) Class 0; (b) Class 1; (c) Class 2; (d) Class 3. Magnification 1000x. (KOBAYASHI et al., 1995).

Studies on damage to DNA integrity in elderly populations are still scarce, but the Comet Assay has shown that the rates of genomic damage in the group aged 85 and over are greater than or equal to those of younger elderly individuals. Hyland et al. (2002) investigated the levels of DNA damage in 21 nonagenarians and 5 controls and found no differences in the integrity of this material compared to controls with an average age of 40.7 years. However, the long-lived had a higher antioxidant capacity. Chevanne et al., in 2003 and 2007, showed that the rates of DNA damage in very old individuals were similar to those of younger individuals (18-76 years), *in* both *in vitro* and *in vivo* studies. On the other hand, in a study involving 40 young people, 35 elderly people and 22 older people, damage to genetic material was higher in the latter group, but their repair mechanisms were more efficient (HUMPHREYS et al., 2007). Recently, the work of Franzke et al. (2015) with 105 institutionalized elderly people, aged 65 to 98, showed that physical exercise did not cause an increase in genomic damage among the elderly under training or not, but there was a better antioxidant capacity and resistance to damage caused by hydrogen peroxide.

These possible responses of resistance to oxidative stress suggest that the increase in DNA damage with age is accompanied by the improved efficiency of its repair and antioxidant mechanisms in healthy aging, which play a critical role in the defense of cells that are routinely exposed to reactive oxygen species at sites of inflammation and defense (FRANCESCHI et al., 2007; HUMPHREYS et al., 2007; KING et al., 1997). These positive, genetically controlled artifices seem to prevent or at least minimize the harmful effects of senescence.

Chapter 8 Strategies for promoting healthy ageing

Although population ageing is widely recognized as one of humanity's greatest achievements, it has profound implications for public policies and represents one of the greatest challenges for contemporary health (LIMA-COSTA; VERAS, 2003).

In Brazil, as described above, the elderly represent a sizable and increasingly important population group (IBGE, 2013a). It is known that the diseases that affect this group are generally multiple and last for several years, being the main factor limiting life span, causing human suffering and economic demands (BUDOVSKY et al., 2007). This panorama generates significant interest in devising measures that are capable of delaying or even preventing the onset of these morbidities.

For the World Health Organization (2005), understanding how individuals are ageing and increasing healthy, quality life expectancy has become an essential goal for tackling this biodemographic scenario.

The most recent instruments for promoting the health of the elderly in Brazil are the guidelines published in 2006 by the Pact for Health and the revision of the National Policy for the Health of the Elderly - PNSI (BRASIL, 2006a;b). The health of the elderly, the promotion of healthy ageing and ongoing research are premises that stand out.

With this in mind, the Ministry of Health has published the Strategic Action Plan for Tackling Chronic-Degenerative Diseases in Brazil, from 2011 to 2022. This plan aims to prepare Brazil to tackle and stop CNCDs, which account for around 70% of deaths over the next ten years, by developing and implementing integrated, sustainable and evidence-based actions for their prevention and control, as well as strengthening health services. This plan defines health promotion and comprehensive care for the elderly as guidelines and considers food to be a modifiable risk factor for CVDs (BRASIL, 2011).

From the perspective of health promotion actions, the use of bioactive compounds, such as glucans, polysaccharides produced by bacteria, yeasts, algae, plants and fungi, has been highlighted, as they have biotechnological potential and are promising targets to help prevent and control chronic degenerative diseases (RAMESH; THARANATHAN, 2003).

Glucans

Glucans are polymers, widely distributed in nature, made up of D-glycopyranoside units (d-Glcp), linked by glycosidic bridges and classified into two groups, α or β -D-glucans, according to the anomeric configuration of the main chain (CHEN; SEVIOUR, 2007; SILVA et al., 2006).

Despite the simplicity of the monosaccharide composition, a wide diversity of these polysaccharides can be found due to the number and configuration of the anomeric carbon of the d-Glpc units, the position and sequence of the glycosidic bonds in the chain, as well as the frequency of branching and the conformation of the chain (SYNYTSYA; NOVAK, 2013).

These biopolymers are produced by bacteria, algae, plants and fungi (MCINTOSH; STONE; STANISICH, 2005) and α- and β-D-glucans can be distinguished by their stereochemical properties. α-glycosidic bonds are formed in the axial position of the chain and are present in the polysaccharides pullulan and dextran, produced by fungi and bacteria. On the other hand, β interactions assume an equatorial position in the chain and are found in the cell wall or extracellular space of fungi and yeasts such as xanthan and scleroglucan or can be obtained from cereals (BARBOSA et al., 2003; BASHARI et al., 2013; SPATAREANU et al., 2014).

There are few records of α-glucans in the literature, but natural products containing β-glucans have been used for thousands of years for their benefits to human health, and their identification as an active component was only described a few decades ago (LUCAS et al., 1958; SHA et al., 2000; WILLIAMS; DI LUZIO, 1980). Since this discovery, these biopolymers, mainly from fungi, have been isolated and extensively studied for their possible biotechnological applications (CHEN; SEVIOUR 2007; WASSER, 2002; ZONG; CAO, WANG, 2012).

Fungi can produce β-glucans of different types, with peculiar physical, chemical, rheological and bioactive properties, located in the cell wall and in the extracellular medium. The latter are integrated into the exterior of the cell wall, forming capsules around the cell or accumulating in large quantities, diffusing into the extracellular medium, and are called exopolysaccharides (EPS). These fungal polysaccharides account for a significant percentage of biomass, contributing more than 75% of the polysaccharides in the hyphae wall (GUTIÉRREZ; PIETRO; MARTINEZ, 1996; SELBMAN et al., 2002).

EPS are related to phytotoxic processes and plant diseases or to plant-pathogen interactions, as elicitors (CORSARO et al., 1998; SELBMANN; SITNGELE; PETRUCCIOLI, 2003). Studies have shown their participation in the degradation of lignin, as an indirect source of hydrogen peroxide, adsorption of excreted enzymes, protection of hyphae against dehydration and regulation of the concentration of extracellular glucose (BUCHALA; LEISOLA, 1987; KRCMAR et al., 1999).

In the food industry, these β-glucans have a wide range of applications due to their thickening, stabilizing, emulsifying and gelling properties (AHMAD et al., 2012).

However, considering their potential as a biological response modifier (BRM) and their use on a commercial scale, EPS have also been gaining attention in recent years. In addition to immunoprotective activity, antithrombotic, antioxidant, antiviral, anti-inflammatory and antiproliferative actions, as well as effects on risk factors for cardiovascular diseases, diabetes and infections have been described for these polysaccharides (BRANDI et al., 2011; CUNHA et al., 2012; KATO et al., 2010; MARTINICHEN- HERRERO et al., 2005; SOLTANIAN et al., 2009; SYNYTSYA; NOVAK, 2013; WANG et al., 2010). In addition to these characteristics, these biopolymers are secreted into the growth medium, are easily recovered through alcohol precipitation, followed by purification and chemical characterization and are not susceptible to climate change, ocean contamination or harvesting adversities, as in the case of those of plant or marine origin (CHEN; SEVIOUR, 2007; VASCONCELOS et al., 2013).

These numerous biological activities result in a multitude of applications for these EPS in the food, cosmetics, pharmaceutical and biomedical sectors.

Exopolysaccharide Botriosferana

The ascomycete and lignolytic fungus, *Botryosphaeria rhodina* MAMB-05, was isolated from eucalyptus canker and produces an exopolysaccharide described as a β-D-glucan (BARBOSA; DEKKER; HARDY, 1996; DEKKER; BARBOSA, 2001).

Subsequently, Barbosa et al. (2003) characterized this polymer structurally as an EPS ((1→3^→6)-β-D-glucan), which was named botriosferana. This polysaccharide is made up of a main, linear chain of glucose residues (homopolymer), joined by β(1→3) glycosidic bonds, with glucose and gentiobiose (two glucose molecules) branching off every five carbon residues of the chain, by β(1→6) bonds. The chemical structure of this β-glucan is shown in figure 8.

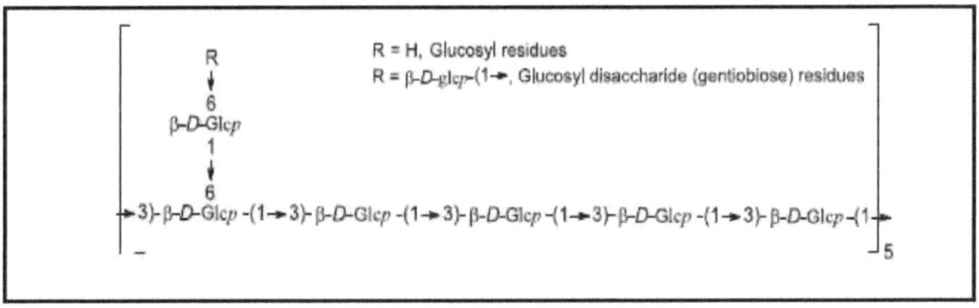

Figure 8. Chemical structure of (1→3^→6)-β-D-glucan from the exopolysaccharide botriosferana (BARBOSA et al., 2003).

Many microorganisms produce these polysaccharides in response to environmental factors, such as the carbon source, which determines the quantity and quality of the EPS obtained (ROSADO et al., 2003). When evaluating possible carbon sources for the production of botriosferana, Steluti et al. (2004) observed that *Botryosphaeria rhodina* synthesized this EPS from most of the carbohydrate monomers and dimers used (glucose, fructose, galactose, mannose, sorbitol, lactose, sucrose, commercial sucrose and sugar cane molasses). However, commercial sucrose was the best carbon source, with the highest production of botriosferan. In addition, the use of this raw material in the composition of the growing medium appears to be quite feasible, considering its cost and easy accessibility, as well as the commercial-scale production of this EPS.

In addition, this homopolymer has considerable solubility in water, being able to form gels or viscous solutions, which favors its use in the food, pharmaceutical and chemical industries (NOVAK; VETVICKA, 2008; KAGIMURA et al., 2015).

Several studies have demonstrated the bioactivity of this EPS, which places it in the group of compounds that modify biological responses. Miranda et al. (2008) showed that botriosferan was not mutagenic and had a chemoprotective effect *in an in vivo* test. In a study carried out with *Sprague-Dawley* rats and RAW264.7 macrophage cells, this biopolymer was found to have a potent immunomodulatory activity (WENG et al., 2011). With regard to possible activity on biochemical parameters, Miranda-Nantes et al. (2011) found a considerable hypoglycemic effect of this polysaccharide in *an in vivo* diabetes induction trial. More recently, its antioxidant, antiproliferative and pro-apoptotic activities were confirmed *in in vitro* trials with MCF-7 breast cancer cells and Jurkat human leukemic lymphocytes (GIESE et al., 2015; MALINI et al., 2015; QUEIROZ et al., 2015).

The mechanisms involved in the biological responses of β-glucans have yet to be fully elucidated. *In in vivo* trials, it is believed that a cascade of events occurs with the activation of macrophages and their receptors, followed by the activation of signaling pathways for the expression of genes that regulate apoptosis during cell proliferation and invasion (NOVAK; VETVICKA, 2009). Furthermore, in vertebrates, the modulation of the immune system by these EPS is due to their contact with receptors and their ability to activate leukocytes and cause a pro-inflammatory state (SYNYTSYA; NOVAK, 2013).

These fungal β-glucans have been used by Japanese and Chinese medicine, traditionally and for millennia, to treat chronic degenerative diseases such as diabetes, neurodegenerative and cardiovascular diseases, as well as to increase lifespan (BORCHERS et al., 1999; ZONG; CAO; WANG, 2012). Its effectiveness and possible biological and/or physiological pathways have also been discussed. Its antihyperlipidemic and antihyperglycemic effects are believed to be related to its ability to interfere with the intestinal absorption of cholesterol and glucose, as the main constituents of dietary fiber, to stimulate the production of bile acids and to influence the synthesis of the hormone insulin (CASTRO; TIRAPEGUI; BENEDICTO, 2003; EL KHOURY et al., 2012). In this case, it seems that the bioactivity of these polymers is related to their molecular structure, including their monosaccharide composition, glycosidic bonds in the main chain and degree of branching (FREITAS et al., 2009; NOVAK, VETVICKA, 2008).

Despite the vast interest and research already carried out by our group, there is still much to be explored regarding the therapeutic potential and biotechnological applications of EPS ((1→3^→6)-β-D-glucan). To date, research has been carried out *in vitro* or with adult mice. Studies aimed at investigating the (anti)cytogenotoxic effect and bioactivity of botriosferan in male and female *Swiss* mice (*Mus musculus*), young and old, subjected to subchronic treatment and in an animal model with gene ablation of a chronic degenerative disease are important because of their role in expanding scientific knowledge and the biotechnological potential of this biomacromolecule. The data generated could help to increase the evaluation of its safety in terms of cytogenotoxicity, at different stages of the life cycle, and provide subsidies for patent registration of a new food ingredient or product for biotechnological purposes, as an adjuvant in the prevention or treatment of diseases that accompany ageing.

Chapter 9 Animal Tests

Despite the progress observed in the use of alternative methods, such as *in vitro* tests, experimental models with animals continue to be worthy of merit, mainly because they provide comprehensive information on the whole organism (CHIA et al., 2005; RIBEIRO; CAMPOS; TIRAPEGUI, 1995).

Among *in vivo* models, the use of mice has stood out since the beginning of the 20th century. The choice of mice for scientific research is due to their small size, high proliferation, short gestation period, easy domestication and maintenance (SANTOS, 2002a;b). In addition, it has a considerable number of established experimental protocols and makes up a multitude of existing models (LEON, 2005).

Murines are also the animals of choice for evaluating the genotoxic potential of agents with biotechnological applications, such as botriosferan, since previous tests for evaluating and approving new products using rodents are recommended by the main regulatory bodies (ANVISA, 2013a;b; CSGMT, 1988).

According to Richardson et al. (2016), when it comes to biogerontology research, mice are the mammalian model of choice. In addition to the characteristics described above, these animals are interesting for studying ageing and the diseases that accompany it, due to their relative ease of genetic manipulation, low cost and short lifespan.

According to the same author, genetic characteristics and sex can have a significant effect on ageing, as well as on physiological functions and diseases. It is therefore imperative that studies be carried out on animals of both sexes, with genetic heterogeneity, and, if possible, also on isogenic strains.

Mutagenicity and antimutagenicity assessment

As described above, exopolysaccharides such as β-D-glucans have aroused great interest due to their technological properties and biological activities. With regard to botriosferan, studies involving its efficacy as a bioactive compound and its toxicological safety have been carried out by our research group. Although it has been extensively studied, its mutagenic, cytotoxic and antimutagenic potential *in in vivo* tests on young and old animals has not yet been explored.

According to national and international regulatory agencies, assessing the genotoxic hazard of compounds with therapeutic potential is an important stage in pre-clinical safety studies (ANVISA, 2013a;b; SAWANT; FIELDEN; BLACK, 2014).

DNA damage can manifest itself in various ways, such as point mutations, double or single strand breaks, adduct formation and recombination events (KIRSCH-VOLDERS et al., 2002). A battery of *in vitro* and *in vivo* tests has been adopted by Toxicological Genetics to evaluate the mutagenicity potential of new bioactive agents (OECD, 1997).

Among these tests, the *in vivo* Micronucleus Test has been widely accepted and recommended in the investigation of clastogenic agents, which break chromosomes, and aneugenic agents, which induce aneuploidy or abnormal chromosome segregation (MACGREGOR et al., 1987; KRISHNA; HAYASHI, 2000). This test evaluates the formation of micronuclei (MN) in samples of erythrocytes from the bone marrow or peripheral blood of rodents, which is especially relevant because it is fast, low-cost, sensitive and takes into account processes such as the metabolism of the test substance, its pharmacokinetics and DNA repair, which are active and contribute to distinct responses (OECD, 1997; KRISHNA; HAYASHI, 2000; TWEATS et al., 2007).

The initial proposal for this cytogenetic test was developed at the end of the 20th century, and the micronuclei were called "Howell-Jolly bodies" (SCHMID, 1975; SEARS; UDDEN, 2011). Subsequently, Heddle et al. (1983) modified the method, from its design and protocols to statistical analysis, and attested that the micronucleus was a sensitive biomarker of genotoxicity.

MN are a small mass of chromatin or tiny nuclear bodies, separated from the main nucleus during telophase by a delay during migration to the poles of the cell in the anaphase of cell division, originating from acentric chromosome fragments, chromosome breaks or whole chromosomes (Figure 9). Specifically, during the process of erythropoiesis, the main nucleus is expelled from an erythroblast and the MN, a fragment or the entire chromosome remains in the cytoplasm, surrounded by the nuclear membrane, in the immature erythrocyte and can be visualized by cytological analysis (SEDELNIKOVA et al., 2007; FENECH et al., 2011).

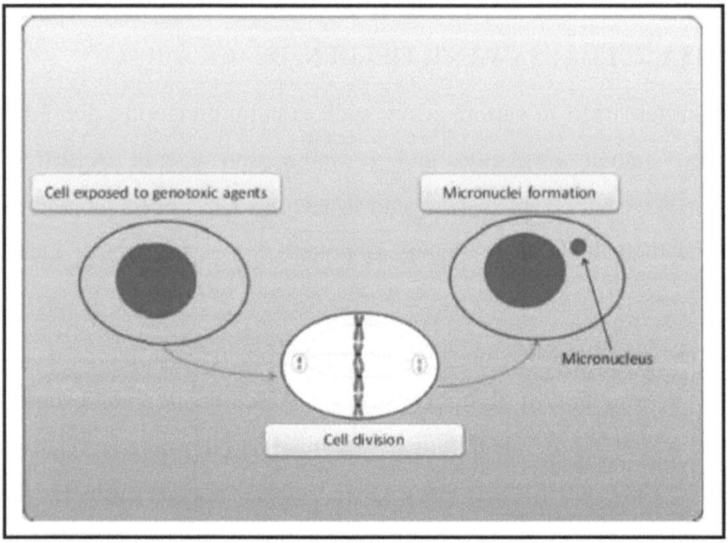

Figure 9. Formation of micronuclei in cells exposed to genotoxic agents (LUZHNA; KATHIRIA; KOVALCHUK, 2013).

According to Ribeiro (2003), during bone marrow analysis, the effect of the potentially toxic agent is observed in these immature cells, called anucleated polychromatic erythrocytes (PCE), which have a relatively short half-life of up to twenty-four hours, which ensures that the detection of any micronuclei generated is the result of recently induced chromosomal damage. The MN are easily identifiable, their distribution being well defined, and the PCE still contain ribonucleic acid (RNA) and stain blue in the presence of Leishman's follow-up methylene blue eosin. Mature normochromatic erythrocytes (NCE) are also identified with a pink color when the same dye is present, since they no longer contain RNA, but hemoglobin, in a greater proportion (KRISHINA; HAYASHI, 2000). In the end, it is possible to estimate the mutagenic potential of the test substance by calculating the PCE and MN between control and treatment materials, and with the ratio PCE: (PCE+NCE), its cytotoxicity.

In mice, these sub-chronic or chronic toxicity tests can also be carried out on peripheral blood, by analyzing micronuclei in NCE, since the spleen of these animals does not remove erythrocytes with micronuclei from the bloodstream. This aspect is appropriate because more than one sample from the same animal can be obtained, without the need for euthanasia, which makes it possible to assess the frequency profile of micronuclei over the course of a treatment (CSGMT, 1992; SCHLEGEL; MACGREGOR, 1982). Hayashi et al. (1990) reported on the micronucleus test in peripheral blood, using acridine orange dye. This fluorescent dye flows rapidly through the cytoplasm, across the cell membrane and binds to DNA and RNA with high affinity (BENVIN et al., 2007). However, it has the property of intercalating with DNA, emitting yellow fluorescence, which is not observed when it binds to RNA, generating red fluorescence. This distinct ability makes it possible to identify normochromatic erythrocytes, which are rich in RNA, and micronuclei, which contain DNA.

Although they can originate spontaneously, the induction of micronuclei can be caused by the use of agents that produce a substantial increase, reliably detected by their frequency, higher than the modest levels normally found (OECD, 1997). These substances are administered to the positive controls during the tests, with cyclophosphamide being one of the agents chosen for this purpose. Cyclophosphamide, used as an antineoplastic, is an alkylating agent of the nitrogen mustard type, which interferes with DNA synthesis, its main pharmacological action being related to covalent binding to nucleophilic compounds in dividing cells. This alkylation process generates significant clastogenic effects (COLVIN, 1999; KADAM; MOHADIK; BOTHARA, 2007).

In addition to exploring the genotoxic danger of a bioactive compound, it is also possible to evaluate its antimutagenic or chemoprotective potential using the Micronucleus Assay. In this proposal, treatments with mutagenic and antimutagenic agents can take place simultaneously, before or after the induction of DNA damage (ANTUNES; ARAÙJO, 2000).

The term "antimutagenic" was first used by Novick and Szilard in 1952 and represents the peculiar characteristic of agents that reduce the frequency of spontaneous or induced mutations, regardless of the mechanism involved (VON BORSTEL; DRAKE; LOEB, 1996). Two mechanisms are proposed for these protective agents: demutagenesis and bioantimutagenesis. In the former, antimutagenic agents are able to inactivate, by direct action, the compounds that induce mutations, chemically or enzymatically, by inhibiting the metabolic activation of pro-mutagens or by sequestering reactive molecules. In bioantimutagenesis, protective agents act on the process that induces mutations, as modulators of DNA repair and replication (FERGUSON, 1994; KADA; MORITA; INOUE, 1978; WATERS et al., 1996).

Studies have shown that β-D-glucans have antitumor, anticancer and anti-inflammatory activities, which makes them promising therapeutic agents (CHEN; SEVIOUR 2007; MANTOVANI et al., 2008; PETRAVIC-TOMINAC et al., 2010; SURENJAV et al., 2006). Miranda et al. (2008) found that botriosferan showed dose-dependent antimutagenic activity against the clastogenic effects induced by cyclophosphamide, but did not investigate its cytotoxicity. With regard to the possible mechanisms involved *in* this biological effect, Giese et al. (2015) found that this EPS exhibited a high natural antioxidant capacity *in an in vitro* test for eliminating reactive oxygen and nitrogen species, as well as considerable elimination of these unstable species. *In* addition, Queiroz et al. (2015) and Malini et al. (2015) showed, in *in vitro* tests with MCF-7 breast cancer cells and Jurkat human leukemia lymphocytes, significant antiproliferative activity, as well as gene modulation by botriosferan, which give it important biomodulatory properties in the treatment of cancer.

In this sense, this (1→3^→6)-β-D-glucan could protect cells against the effects produced by genotoxic agents, making it possible in the future to reduce the risk of cancer and other chronic-degenerative diseases that affect the elderly population.

It is known that the aging process and DCDs are accompanied and/or caused by increased oxidative damage to cells and genetic material (KOHEN; NISKA 2002; DATO et al., 2013). The Micronucleus Test has been considered one of the main markers of genomic instability (LUZHNA; KATHIRIA; KOVALCHUK, 2013), and it is possible, through parameters such as antimutagenicity and anticytotoxicity, in trials with elderly mice, to establish innovative and promising strategies in the use of biomolecules such as botriosferan to mitigate the effects of these oxidative lesions.

Having achieved satisfactory results regarding the cytogenotoxic safety of botriosferan exopolysaccharide, studies involving its efficacy with regard to the prevention and treatment of chronic degenerative diseases, such as cardiovascular diseases, are extremely important, given its impact on the costs of health systems and on the quality of life of the world's population.

Animal Model for Atherosclerosis

Cardiovascular diseases (CVD) are considered the leading cause of death in the world, accounting for more than 17.3 million deaths annually (WHO, 2011). In 2011, 335,213 people in Brazil died from CVD, 37.17% of whom were aged 60 or over. In addition, these diseases were the second leading cause of heart attacks, accounting for 30% of all deaths (DATASUS, 2011).

Atherosclerosis is a chronic degenerative disease, considered to be the primary cause of other cardiovascular diseases, such as stroke and ischemic heart disease, due to endothelial aggression that mainly affects the intimal layer of medium and large caliber arteries (HOOG et al., 1993; LUSIS, 2000). The atherosclerotic process is progressive and slow, characterized by the accumulation of lipids, inflammatory cells and fibrous elements, and its etiology is multifactorial with a combination of genetic and environmental factors and life span (HACKAM; ANAND, 2003; LIBBY, 2002).

The hypothesis that low-density lipoprotein (LDL-c), which carries 65% of the cholesterol in the bloodstream, is the one that plays a relevant role in the pathogenesis of atherosclerosis, in relation to others (intermediate-density lipoprotein -IDL, very low-density lipoprotein -VLDL or chylomicrons), is one of the most widely accepted. Endothelial aggression, caused by various factors such as hypercholesterolemia, caused by a higher intake of cholesterol from the diet or by genetic variants that lead to a decrease/alteration of LDL receptors, increases plasma concentrations of LDL-c. Endothelial dysfunction increases the permeability of the intimal layer of the vessels to lipoproteins, favoring their retention in the subendothelial space. LDL-c is oxidized by reactive oxygen species, becomes immunogenic and endothelial damage is formed. In the endothelium, adhesion molecules are expressed and inflammatory response mediators are produced, with the presence of macrophages, filled with lipids, forming fatty streaks, macroscopic lesions, and T lymphocytes, which produce cytokines (HANSSON, 2005; O'BRIEN; NGUYEN, 1997; ROSS, 1999; STEINBERG et al., 1989). Intense inflammatory activity, accompanied by considerably thrombogenic material, can culminate in the formation of an underlying thrombus, a process called atherothrombosis, which is responsible for some of the clinical manifestations of atherosclerosis (XAVIER et al., 2013).

The formation of atherosclerotic lesions is a slow, progressive process and can take decades to develop in humans (GOTTLIEB; BONARDI; MORIGUCHI, 2005). However, murine models have been established and are being widely used. The pathological process is very similar, occurs over a short period of time and the experimental models stand out due to their easy handling, reproduction and availability of isogenic strains (VANDERLAAN; REARDON; GETZ, 2004; VASQUEZ et al., 2012).

In the 1990s, genetic engineering developed mice to be used as models for studying experimental hypercholesterolemia. Ishibashi et al. (1993) reproduced the phenotype of familial hypercholesterolemia using homologous recombination in embryonic cells, producing mice with the LDL receptor gene inactivated ($LDLr^{-/-}$). The allele-specific gene deletion generated *knockout* mice with plasma cholesterol levels twice as high as those observed in wild-type or control mice (C57BL/6 - isogenic *background* strain), due to an increase in the atherogenic lipoproteins LDL and IDL. However, the development of atherosclerotic lesions became more significant when these animals were fed an atherogenic diet, with a high content of saturated fat and cholesterol, also called *Western-type* (similar to the Western diet, with a significant contribution of lipids).

According to Bergman and Ader (2000), there is a significant relationship between lipid metabolism disorders and insulin resistance. Thus, experimental models with β-pancreatic cells isolated from mice and human islets have shown that oxidized LDL plays a relevant role in the pathogenesis of insulin resistance, impairing the survival and secretory function of beta cells (ROEHRICH et al., 2003; RUTTI et al., 2009). In the study by Bonfleur et al. (2010), LDLr $/^{-}$ mice showed impaired glucose tolerance and insulin secretion. The islets of these animals had a significant cholesterol content and their mitochondria produced a greater amount of reactive oxygen species. This oxidative state favors the transition of membrane permeability, which culminates in cell death. Insulin resistance can also be attributed to the increase in visceral adipose tissue caused by the high-fat diet (KOWALTOWSKI; CASTILHO; VERCESI, 2001; OLIVEIRA et al., 2005).

Among the advantages of using this animal model, we can mention the similarity to the phenotype of human familial hypercholesterolemia, which results from mutations in the gene for the LDL receptor; the fact that most of the cholesterol is confined to low-density lipoprotein (LDL), which makes the perii! The degree of dyslipidemia is intermediate and the lesions are less advanced when compared to models with gene ablation for apolipoprotein E, which acts to purify chylomicrons and remaining VLDL (ZADELAAR et al., 2007).

It is important to note that in mice, the fraction responsible for the highest percentage of cholesterol transport is not LDL-c, but HDL-c (High Density Lipoprotein), which explains the resistance of wild animals to developing atherosclerosis. Despite this difference, mice generally have the same set of genes that modulate our lipid metabolism and the lipid profile of animals developed with gene deletion is similar to that observed in humans, with a shift in plasma lipoproteins from HDL-c to VLDL-c, IDL-c and LDL-c (BRESLOW, 1996; JAWIEN; NASTALEK; KORBUT, 2004).

Effects of β-glucans on glycemia, lipid profile and atheroprotection

Animal models of human diseases are of great relevance in studies that seek to investigate the possible therapeutic effects of new chemical entities such as botriosferan. Associated with this information, data from the literature indicate that β-glucans are effective in preventing and treating different pathological conditions and their complications (KAGIMURA et al., 2015; ZHU, DU, XU, 2016).

The biological effects of β-glucans from cereals such as oats have been widely explored scientifically and are already regulated, generally recognized as safe (GRAS), being admitted as the predominant constituents of soluble fibers with the ability to modify health indicators related to DCDs such as diabetes and hypercholesterolemia (LAZARIDOU; BILIADERIS, 2007; LAZARIDOU et al., 2014; US FDA, 1997).

In this sense, Watkins (2004) suggested the inclusion of β-glucans in the diet, as they contribute to the homeostasis of cholesterol and glucose metabolism, as an adjuvant agent in the treatment of hypercholesterolemia, which can lead to other cardiovascular diseases.

In yeasts, Waszkiewicz-Robak (2013) showed this pro-health activity when they offered β-glucan from *Saccharomyces cerevisiae* to mice that had received a high-fat diet. Vetvicka and Vetvickova (2007), in the first study to induce hypercholesterolemia through a high-fat diet, with mice as the experimental model, evaluated the cardioprotective efficacy of β-glucans isolated from cereals, yeast and fungi, and observed their hypocholesterolemic effect. London et al. (2014), analyzing the antilipidemic potential of β-glucans produced by bacteria in a trial involving young mice with a gene deletion for apolipoprotein E, found that those on an atherogenic diet in the presence of these biopolymers showed a reduction in hypercholesterolemia and hypertriglyceridemia, but no atheroprotective effect on atherosclerotic plaques.

As previously described, these biopolymers have also been isolated from the cell wall or extracellular medium of fungi. The effectiveness of a β-glucan, extracted from the fruiting body of *Volvariella volvacea,* in reducing the levels of total and hepatic cholesterol, LDL-c and plasma triglycerides in hypercholesterolemic rats was reported in the study by Cheung (1996). In addition, this glucan promoted an increase in the excretion of bile acids. Using a similar experimental model, Yang et al. (2002) also showed the hypolipidemic activity of an EPS from *Auricularia polytricha*. In a study with rats to induce diabetes, fungal polysaccharides were able to decrease plasma levels of glucose, cholesterol and triglycerides (KIM et al., 2005). In 2007, Cho et al. showed that EPS extracted from *Tremella fuciformis and Phellinus baumii* had an antihyperglycemic and hypotriglyceridemic effect when supplemented orally to mice for fifty-two days. The exopolysaccharide isolated from *Stropharia rugosoannulata* also showed an antihyperglycemic effect *in an in vivo* trial *(*ZHAI et al., 2013). In young mice given a high-fat diet, the hypocholesterolemic activity of *Ganoderma licidum* was observed (RUBEL et al., 2011). Similarly, the antihyperglycemic, antihyperlipidemic and antioxidant activities of polysaccharides extracted from *Catathelasma ventricosum* were observed in mice treated for thirty days as an experimental model of diabetes (LIU et al., 2013).

Our research group demonstrated that botriosferan, at a dose of 12mg/kg of body weight, showed hypoglycemic activity during a fifteen-day treatment in a diabetes induction trial with adult rats and less significant benefits in hypercholesterolemia in rats given a high-fat diet (MIRANDA-NANTES et al., 2011).

However, experimental models of chronic degenerative diseases such as atherosclerosis, using elderly animals, exploring the biotechnological potential of this exopolysaccharide (1→3^→6)-β-D-glucan have not yet been carried out.

As well as in animals, research in humans has also shown the benefits of eating these polysaccharides. Hypercholesterolemic individuals aged between 27 and 78 showed lower plasma cholesterol levels when given three grams (g) of glucans daily as food ingredients (WANG et al., 2016). Among patients with type 2 *diabetes mellitus*, *Pleurotus ostreatus* was also promising in lowering blood glucose and significantly improving the lipid profile after three months of intake (AGRAWAL et al., 2010). Children and adolescents diagnosed with familial hypercholesterolemia, when given 3g of β-glucans, as an ingredient in food products and isolated from *Sacharomyces cerevisiae***,** in their diet for four weeks, showed a change in their lipid profile, with a reduction in total cholesterol (22%) and low-density lipoprotein (28%) levels (MAKI et al., 2003).

The literature on the positive effects of β-glucans on health is considerably extensive. However, the possible cellular and molecular mechanisms involved are still being debated.

These exopolysaccharides, when ingested, pass through the stomach without undergoing chemical or structural changes because they are resistant to acidic media and, in the intestine, the alkaline hydrogen potential favors the ionization of hydroxyl groups and their solubility, as well as there being no enzymes capable of degrading them (BER, 1997). According to Castro, Tirapegui and Benedicto (2003), these biomacromolecules could interfere with the intestinal absorption of cholesterol and the production of bile acids. Chema et al. (2016) described that fungal EPS can act as prebiotics and their fermentation in the colon culminates in the production of short-chain fatty acids, which are involved in the hepatic regulation of lipids. The hypoglycemic effect is related to a decrease in the levels of hepatic and pancreatic transaminases, which increases the production of the hormone insulin (KIM et al., 2001; VETVICKA; VETVICKOVA, 2007). El Khoury et al. (2012) also reported the effects of β-glucans on reducing biochemical parameters such as cholesterol and glucose, and associated them with their ability, as a soluble fiber, to increase viscosity in the intestine, delaying gastric emptying, digestion and absorption. This reduces the transport of these compounds to the enterocytes.

According to the hypothesis of Kahlon and Woodruff (2003), in the intestine, these biopolymers would bind to bile acids, which would increase their excretion in the feces. This process would result in a decrease in the *pool of* bile salts (BS) active in the formation of micelles, an important step in the absorption of lipids, and would interfere with their synthesis from cholesterol in the liver. In this way, both the impaired transport of cholesterol into the bloodstream and its mobilization to produce SB would lead to a reduction in its plasma concentration. These findings were also described by Bobek, Ozdin and Galbav (1998), in an in *vivo* trial to induce hypercholesterolemia.

Fukushima et al. (2000) suggested that dietary fibers, especially those rich in β-glucans, lowered total serum cholesterol by increasing the expression levels of messenger RNA for the hepatic LDL receptor *in an in vivo* trial. The results of Hong, Xun and Wutong (2007) showed that the anitidiabetic effect of glucans from *Grifola frondosa*, in an experimental model with mice, was due to their influence on increasing insulin sensitivity and reducing insulin resistance. *In* relation to this assumption, *in* an *in vivo* trial with *Ganoderma lucidum* polysaccharides, Xiao et al. (2012) found that the decrease in plasma glucose levels was associated with a reduction in the expression levels of enzymes involved in gluconeogenesis and/or glycogenolysis.

As for the possible production of short-chain fatty acids by β-glucans, Russo et al. (2012) proved the prebiotic effect of an exopolysaccharide from *Pediococcus parvulus* for three different probiotic strains. *In* addition, the presence of this EPS enhanced the adhesion of *Lactobacillus plantarum* WCFS1 to intestinal epithelium cells *in an in vitro* test. A prebiotic is defined as a food ingredient that cannot be digested but stimulates the growth of microorganisms that are beneficial to our intestinal microbiota (GIBSON et al., 2004). These microorganisms can produce short-chain fatty acids, resulting from the fermentation of the prebiotic, which improve the lipid profile by inhibiting the activity of the enzyme hydroxymethylglutaryl coenzyme reductase and, therefore, the synthesis of hepatic cholesterol, and/or its redistribution from the plasma to the liver (PEREIRA; GIBSON, 2002). It is speculated that these polysaccharides may inhibit the expression of lipogenic enzyme transcripts through cell receptors and secondary mediators, which would imply a reduction in circulating VLDL-c (KAUR, GUPTA, 2002).

More recently, there has been discussion about the effect of β-glucans on the set of signs and symptoms of metabolic syndrome, which are related to alterations in parameters of lipid and glucose metabolism. Cloetens et al. (2012) summarized the latest human studies on the subject and showed that the increase in chyme viscosity caused by the ingestion of these glucans during the digestive process seems to play a crucial role in reducing blood glucose. It was concluded that the molecular structure of these biomolecules seems to have an impact on plasma glucose levels and the response to insulin, which reinforces the importance of knowing the chemical and structural composition of these polysaccharides in order to target their possible therapeutic effects. Legentil et al. (2015) considered that the molecular weight, degree of branching, solubility, three-dimensional conformation and the way in which glucans interact with their receptors are crucial for them to exert their biological effects. With regard to botriosferan, its main chain is linear, but with branching, which favors its solubility. In addition, its triple helix conformation enables the formation of interchain hydrogen bonds, which contributes to the molecule's stability (GIESE et al., 2008).

Notwithstanding the conjecture of Raa (2015), who describes that the biological activity of (1→3^→6)-β-D-glucans seems to occur at the intestinal level and their possible systemic effects result from this primary interaction, via membrane receptors and secondary mediators, the existence of cellular receptors and binding proteins in plasma specific to β-glucans in humans has been described by Chan, Chan, Sze (2009) and Ross et al. (1999). These receptors have been identified in cells of the immune system, such as monocytes, macrophages and neutrophils, and in the cells of the alveolar epithelium, endothelial cells and fibroblasts, through *in vitro* studies (BROWN; GORDON, 2003). A small amount of soluble β-glucans is recognized via transmembrane receptors called Dectin-1 in enterocytes and captured by macrophages (BROWN; GORDON, 2001). They are internalized and fragmented inside these cells, generating smaller fragments. Considering that Yadomae (2000) identified undegraded β-glucans in hepatocytes and the spleen, it is suggested that the former or those fragments are eventually secreted by macrophages into the bloodstream and recognized via other receptors, such as placental collectin-1 (CL-P1), expressed mainly in vascular endothelial cells. This type II transmembrane protein can also bind to oxidized LDLs, participating in their internalization by endocytosis and therefore being involved in the atherosclerosis process (CHAN; CHAN, SZE, 2009; JANG et al., 2014; MORI et al., 2014).

The transport described above may support the assumption that the cardioprotective effect of botriosferan is not restricted to the intestinal tract, considering its activity in reducing atherosclerotic plaques. When internalized by endothelial cells, botriosferan acts as an antioxidant agent. Studies have shown that antioxidants can slow down the development of atherosclerotic plaques *in in vivo* trials and it is known that the oxidation of LDL-c plays an important role in atherogenesis (JEMAI et al., 2008; KOROU et al., 2010). (WHITMANN et al., 2004; ROSS, 1999). This biological effect of botriosferan was reported by Giese et al. (2015) who showed the significant antioxidant capacity and ability to eliminate reactive oxygen species of this EPS. This could be an important therapeutic strategy in the treatment and prevention of atherosclerosis.

Another point worth highlighting is the possibility of β-glucans having pleiotropic effects. In this case, the activity of macrophages, which are present in the process of atherosclerotic plaque development, would be modulated by EPS, with the production of lower levels of nitric oxide (NO) and, consequently, peroxynitrite, which would attenuate the inflammatory process characteristic of this pathological condition (BUTTERY et al., 1996). Macrophages from the peritoneal region of hypercholesterolemic mice that received β-glucans from *Ganoderma lucidum* showed lower NO concentrations (RUBEL et al., 2011). Normally, NO plays a protective role as an antioxidant and vasodilator, but at higher levels, induced by pathological conditions and cytokines, it is related to endothelial dysfunction, lipoprotein oxidation, hypercholesterolemia and increased susceptibility to the development of atherosclerosis (NAPOLI; IGNARRO, 2009). It is suggested that this immunomodulatory activity of EPS (1→3^→6)-β-D-glucan occurs via nuclear receptor-signaling to decrease the transcripts of the enzyme inducible nitric oxide synthase (iNOS).

In view of the above, studies into toxicological safety and chemoprotection, as well as the bioactivity of compounds with functional and/or health properties, such as botriosferan, are of the utmost importance in order to provide the necessary information to prove their effectiveness as a promising adjunct, in the form of a food ingredient or medicine, in the treatment and prevention of diseases. These measures are considered a priority in contemporary health, since the elderly population has been increasing its lifespan, but the gains in healthy lifespan have been negligible, as described above. Combining personalized medicine and genomics strategies with the development of products from natural sources with biotechnological potential is establishing emerging ways of intervening in health, with a view to tackling chronic-degenerative diseases and longevity in clinical practice.

BIBLIOGRAPHICAL REFERENCES

AGRAWAL, R.P. et al. Effect of oyster mushroom on glycemia, lipid profile and quality of life in type 2 diabetic patients. **Aust. J. Med. Herbalism**, v. 22, n. 2, p. 50-54, 2010.

AHMAD, A. et al. Beta glucan: a valuable functional ingredient in foods. **Crit Rev Food Sci Nutr**, v. 52, n. 3, p. 201-212, 2012.

ALCENDOR, R.R. et al. Sirt1 regulates aging and resistance to oxidative stress in the heart. **Circ Res**, n.100, p.1512-1521, 2007.

AKBARALY, T. et al. Does overall diet in midlife predict future aging phenotypes? A cohort study. **Am J Med**, v. 126, p. 411-419, 2013.

AMBROSONE, C.B. et al. Manganese superoxide dismutase (MnSOD) genetic polymorphisms, dietary antioxidants and risk of breast cancer. **Cancer Res.**, v. 59, p. 602-606, 1999.

BELENGUER VAREAA, A. et al. Oxidative stress as a predictor of longevity; study of cases and controls **Rev Esp Geriatr Gerontol.**, v. 50, n. 1, p.16-21, 2015.

ANSELMI, C.V. et al. Association of the FOXO3 locus with extreme longevity in a southern Italian centenarian study. **Rejuvenation Res.**, v. 12, n.2, p. 95-104, apr. 2009.

ANTUNES, M.V. et al. Pre-analytical and validation study for the determination of malondialdehyde in human plasma by high performance liquid chromatography, after derivatization with 2,4-dinitrophenylhydrazine. **Rev. Bras. Cienc.**, v.44, n.2, p. 279-287, 2008.

ANTUNES, L.M.G.; ARAUJO, M.C.P. Mutagenicity and antimutagenicity of the main food colorings. **Rev. de Nutr.**, v. 13, n. 2, p. 81-88, 2000.

ANVISA. National Health Surveillance Agency. Guide to the Conduct of Non-clinical Toxicology and Pharmacological Safety Studies Necessary for the Development of Medicines. Safety and Efficacy Evaluation Management. Distrito Federal: Brasilia, January, 2013a, 48p.

ANVISA. NATIONAL HEALTH SURVEILLANCE AGENCY. Guide to Proving the

Safety of Food and Ingredients. Special Products Management and General Food Management. Federal District: Brasilia, February, 2013b, 45p.

ARKING, R. et al.

Identical longevity phenotypes are characterized by different patterns of gene expression and oxidative damage. **Exp Gerontol.** , v.35, n.3, p.353-373, may 2000.

ASSMANN, K.E. et al. A Healthy Dietary Pattern at Midlife, Combined with a Regulated Energy Intake, Is Related to Increased Odds for Healthy Aging. **J Nutr**, v.145, n. 9, p. 2139-45, 2015.

AUGUSTO, O. Free Radicals as Evil to Life. In:. **Free radicals - good, bad and natural.** Sao Paulo: Oficina de Textos - USP, ch. 2, p.38, 2006.

BALTES, P.B; SMITH, J. **New frontiers for the future of aging: from the successful old age of the young elderly to the dilemmas of the Fourth Age and Third Age**. Social Service of Commerce. ST-Gerência de Estudos e Programas da Terceira Idade.

Translated by Anita Liberalesso Neri. Sao Paulo: v.17, n. 36, p.7-31, jun-2006.

BARBOSA, A.M. et al. Structural characterization of Botryosphaeran: a (1→^1→6)-beta-D-glucan produced by the ascomyceteous fungus, Botryosphaeria sp. **Carbohydr Res.**, v. 29, n. 338, supp. 16, p. 1691-1698, Jul. 2003.

BARBOSA, A.M.; DEKKER, R.F.H.; HARDY, G.E.ST. Veratryl alcohol as an inducer of laccase by an ascomycete, Botryosphaeria sp., when screened on the polymeric dye Poly R-478. **Lett. Appl. Microbiol.**, v. 23, p. 93-96, 1996.

BARZILAI, N.; BARTKE, A. Biological approaches to mechanistically understand the healthy life span extension achieved by calorie restriction and modulation of hormones. J **Gerontol A Biol Sci Med Sci.** , v.64, n.2, p.187-191, 2009.

BARZILAI, N.; GABRIELY, I. Genetic Studies Reveal the Role of the Endocrine and Metabolic Systems in Aging. **J Clin Endocrinol Metab**, v.95, n.10, p. 4493-4500, oct. 2010.

BARZILAI, N.et al. The critical role of metabolic pathways in aging. **Diabetes**, v. 61, p.1315-1322, 2012.

BARTKE A. Minireview: role of the growth hormone/insulin-like growth factor system in mammalian aging. **Endocrinology,** v.146, n.9, p. 3718-3723, sep. 2005.

BASHARI, M. et al. Separation and characterization of dextran extracted from deteriorated sugarcane. **Int J Biol Macromol.**, v.59, p. 246-254, aug 2013.

BECKMAN, K.B.; AMES, B.N. The free radical theory of aging matures. **Physiol. Rev**, v.78, p. 547-581, 1998.

BENVIN, A.L. et al. Fluorescent DNA nanotags: supramolecular fluorescent labels based on intercalating dye arrays assembled on nanostructured DNA templates. **J Am Chem Soc.,** v. 21, n. 129, sup. 7,p. 2025-2034, feb. 2007.

BER, L. Yeast-derived β-1,3-glucan: an adjuvant concept. **Am J Natural Med.**, v. 4, p. 21-24, 1997.

BERGMAN, R.N.; ADER, M. Free fatty acids and pathogenesis of type 2 diabetes mellitus. **Trends Endocrinol Metab**., v. 11, p. 351-356, 2000.

BIELSKI, B.H.J.; ARUDI, M.W.; SUTHERLAND, M.W. A study of the reactivity of HO2/O2- with unsaturated fatty acids. **J. Biol. Chem**., v. 258, p. 4759-4761, 1983.

BOBEK, P.; OZDiN, L.; GALBAV, S. Dose- and time-dependent hypocholesterolemic

effect of oyster mushroom (*Pleurotus ostreatus*) in rats. **Nutrition**, v. 14, n.3, p. 282286, 1998.

BOHR, V. Oxidative DNA damage and repair. **Free Radic Biol Méd**, v. 32, p.804812,2002.

BORCHERS, A.T. et al. Mushrooms, tumors, and immunity. **Proc Soc Exp Biol Med.**, v. 221,p. 281-293.1999.

BRANDI, J., et al. Chemical modification of botryosphaeran: Structural characterization and anticoagulant activity of a water-soluble sulfonated (1→3)(1→6)-β-d-glucan. **J Microbiol Biotechnol**, v. 21, p.1036-1042, 2011.

BRAZIL. Ministry of Health. Health Surveillance Secretariat. Strategic action plan for tackling NCDs in Brazil. 2011-2022. DASS. Brasilia: MS, 2011.148 p.

BRAZIL. Ministry of Health. Ordinance No. 399/GM of February 22, 2006.

Announces the 2006 Health Pact. Consolidation of the SUS and approves the Operational Guidelines of the Pact. Brasilia: MS, 2006a.

BRAZIL. Ministry of Health. Ordinance No. 2.528, of October 19, 2006. Approves the National Health Policy for the Elderly. Brasilia: MS, 2006b.

BRESCIANI, G. et al. The Ala16Val MnSOD gene polymorphism modulates oxidative response to exercise. **Clin Biochem**. v. 46, p. 335-40, 2013.

BRESLOW, J.L. Mouse models of atherosclerosis. **Science,** v.272, p.685-688, 1996.

BROER, L . et al. GWAS of Longevity in CHARGE Consortium Confirms *APOE* and *FOXO3* Candidacy. **J Gerontol A Biol Sci Med Sci**, v.70, n. 1, p. 110-118, jan. 2015.

BROOKS-WILSON, A.R. Genetics of healthy aging and longevity. **Human Genetics,** v. *132*, n. 12, p. 1323-1338, 2013.

BROWN, G.D.; GORDON, S. Immune recognition. A new receptor for beta-glucans.

Nature, v. 6, n. 413, supp. 6851, p. 36-37, sep. 2001.

BROWN, G.D.; GORDON, S. Fungal β-glucans and mammalian immunity.

Immunity, v.16, p. 311-315, 2003.

BUCHALA, A. J.; LEISOLA, M. Structure of the β-glucan secreted by

Phanerochaete chrysosporium in continuous culture. **Carbohydr Res.**, v. 165, p. 146-149, 1987.

BUDOVSKY, A. et al. Longevity network: Construction and implications.

Mechanisms of Ageing and Development, v.128, p.117-124, 2007.

BUDOVSKY, A. LongevityMap: a database of human genetic variants associated with longevity**. Trends Genet.,** v. 29, n. 10, p. 559-60, oct. 2013.

BURTNER, C.R.; KENNEDY, B.K. Progeria syndromes and ageing: what is the connection? **Nat Rev Mol Cell Biol.** v. 11, n. 8, p. 567-78, aug. 2010.

BUTTERY,L.D.etal.Inducible nitric oxide synthase is present within human atheroscl erotic lesions and promotes the formation andactivity of peroxynitrite. **Lab Invest.**, v. 75,n. 1,p. 77-85, jul. 1996.

CAMARANO, A.A.; KANSO, S.; LEITÂO E MELLO, J. How do elderly Brazilians live? In: CAMARANO, A.A. **Os Novos Idosos Brasileiros Muito Além Dos 60?.** ch. 1, p. 25-76. Institute for Applied Economic Research. IPEA. Rio de Janeiro. Sep. 2004.

CAMARANO, A.A.; KANSO, S.; LEITÂO E MELLO, J. How far beyond 60 can Brazil's elderly live? In: CAMARANO, A.A. **Os Novos Idosos Brasileiros Muito Além Dos 60?.** cap. 2, p. 77-106. Institute for Applied Economic Research. IPEA. Rio de Janeiro. Sep. 2004.

CASTRO, I.A.; TIRAPEGUI, J.; BENEDICTO, M.L. Effects of diet supplementation with three soluble polysaccharides on serum lipid levels of hypercholesterolemic rats. **Food Chemistry, v.** 80, p. 323-330, 2003.

CASTRILLON, D.H. et al. Suppression of ovarian follicle activation in mice by the transcription factor FOXO3**. Science,** v. 301, p. 215-218, 2003.

CATALANO, R.; BRUCKNER, T; SMITH, K.R. Ambient temperature predicts sex ratios and male longevity. **Proc. Natl Acad. Sci. USA**, v. 105, p. 2244-2247, 2008.

CHAN, G.C.; CHAN, W.K.; SZE, DM.The effects of beta-glucan on human immune and cancer cells. **J Hematol Oncol.**, v. 10, n. 2: 25, jun. 2009.

CHEMA, B. et. al. Structural Characterization, Technological Functionality and Physiological Aspects of Fungal β-D-Glucans: A Review. **Crit Rev Food Sci Nutr**, v. 56, n.10, p. 1746-1752, jul. 2016.

CHEN, J.; SEVIOUR, R. Medicinal importance of fungal beta-(1→3), (1→6)-glucans. **Mycol Res**, v. 111, p. 635-652, 2007.

CHEUNG, P.C. The hypocholesterolemic effect of extracellular polysaccharide from the submerged fermentation of mushroom**. Nutr Res.**, v. 16, p. 1953-1957, 1996.

CHEVANNE, R. et al. Comparative levels of DNA breaks and sensitivity to oxidative stress in aged and senescent human fibroblasts: a distinctive pattern for centenarians. **Biogerontology**, v. 4, p. 97-104, 2003.

CHEVANNE, M. et al. Oxidative DNA damage repair and parp 1 and parp 2 expression in Epstein-Barr virus-immortalized B lymphocyte cells from young subjects, old subjects, and centenarians, **Rejuvenation Res**. v. 10, p.191-204, 2007.

CHIA, R. et al. The origins and uses of mouse outbred stocks. **Nature Genetics**, v. 37, n.11, p. 1181-1186, 2005.

CHO, E.J. et al. Hypoglycemic effects of exopolysaccharides produced by mycelial cultures of two different mushrooms *Tremella fuciformis* and *Phellinus baumii* in ob/ob mice. **Appl. Microbiol.Biotechnol**., v.75,n.6, p. 1257-1265, 2007.

CHRISTENSEN, K.; JOHNSON, T.E.; VAUPEL, J.W. The quest for genetic determinants of human longevity: challenges and insights. Review. **Nat Rev Genet**. v. 7, n. 6, p. 436-48, jun. 2006.

CLOETENS, L., et al. Role of dietary beta-glucans in the prevention of the metabolic syndrome. **Nutr Rev, v.** 70, n. 8, p. 444-58, aug. 2012.

COLLINS, A. R. The Comet Assay for DNA Damage and Repair: Principle, Applications and Limitations. **Molecular Biotechnology**, v. 26, p.249 - 261, 2004.

COLVIN, O.M. An overview of cyclophosphamide development and clinical applications.

Curr Pharm Des, v. 5, p. 555-560, 1999.

CORSARO, M.M. et al. Chemical structure of two phytotoxic exopolysaccharides produced by Phomopsis foeniculi. **Carbohydr Res**., v. 308, n.3-4, p.349-357, jun.1998.

CRUZ, I.B.M. et al. Association among Oxidized LDL Levels, MnSOD, Apolipoprotein and Polymorphisms and Cardiovascular Risk Factors in an South Brazilian Region Population. **Genetics and Molecular Research** (Online), Ribeirao

Preto-SP, v. 4, n.4, p. 691-703, 2005.

CRUZ, I.B.M. Genetics of aging and its impact on human longevity: theories and evidences that help to prevent age-associated diseases. **PAJAR**, v. 2, n.1, p. 3-14, 2014.

CUNHA, M.A.A. et al. Lasiodiplodan, an exocellular (1(6)→d-glucan from Lasiodiplodia theobromae MMPI: Production on glucose, fermentation kinetics, rheology and anti-proliferative activity. **J Ind Microbiol and Biotechnol**, v. 39, p. 110, 2012.

CSGMT. Collaborative Study Group for the Micronucleus Test. Strain differences in the micronucleus test. **Mutat. Res.**, v. 204, p. 307-316, 1988.

CSGMT. Collaborative Study Group for the Micronucleus Test. Micronucleus test with mouse peripheral blood erythrocytes by acridine orange supravital staining: the summary report of the 5th collaborative study by CSGMT/JEMS.MMS. **Mutat. Res.**, v. 278, p. 83-98, 1992.

DA CRUZ, I.B.M. et al. Prevalence of obesity in the long-lived elderly and its association with cardiovascular risk factors and morbidities**. Rev Assoc Méd Bràs, v.** 50, n. 2, p.172-7, 2004.

DAS, H.K. Isolation, characterization, and mapping to chromosome 19 of the human apolipoprotein E gene. **J Biol Chem**, v. 260, n. 10, p. 6240-7, may 1985.

DATASUS. Information Department of the Unified Health System. Ministry of Health. Health Information. Available at:

<http//tabnet.datasus.gov.br/cgi/deftohtm.exe?sim/cnvobt10uf.def>. Accessed on May 10, 2016.

DATO, S. et al. Exploring the role of genetic variability and lifestyle in oxidative stress response for healthy aging and longevity**. Int. J. Mol. Sci**., v. 14, n. 8, p.16443-16472, 2013.

DAVINELLI, S; WILCOXX, D.C.; SCAPAGNINI, G. Extending healthy aging: nutrient sensitive pathway and centenarian populat. **Immun Ageing**, v.9, p.9-15, 2012.

DE MAGALHAES, J.P.; CHURCH, G.M. Analyses of human-chimpanzee orthologous gene pairs to explore evolutionary hypotheses of aging. **Mech Ageing Dev**, v.128, n.5-6, p. 355-64, may-jun. 2007.

DE ONIS, M.; HABICHT, J.P. Anthropometric reference data for international use: recommendations from a World Health Organization Expert Committee. **Am J Clin Nutr**, v. 64, p.650-658, 1996.

DEELEN, J. et al. Genome-wide association meta-analysis of human longevity identifies a novel locus conferring survival beyond 90 years of age. **Hum Mol Genet**. v. 23, p. 4420-32, 2014.

DEKKER, R.F.; BARBOSA, A.M. The effects of aeration and veratryl alcohol on the production of two laccases by the ascomycete Botryosphaeria sp. **Enzyme Microb Technol.**, v.2, n.28, sup. 1, p. 81-88, 2001.

DEL RIO, D.; STEWART, A.J.; PELLEGRINI, N. A review of recent studies on malondialdehyde as toxic molecule and biological marker of oxidative stress. **Nutr Metab Cardiovasc Dis.** , v. 15, n. 4, p. 316-328, aug. 2005.

DE ZWART, L.L. et al. Biomarkers of free radical damage applications in experimental animals and in humans. **Free Radic Biol Med.** v. 26, n.1-2, p. 202-26, jan.1999.

DI BONA, D. Association between genetic variations in the insulin/insulin-like growth factor (Igf-1) signaling pathway and longevity: a systematic review and meta-analysis. **Curr Vasc Pharmacol.**, v.12, n.5, p. 674-681, 2014.

DIZDAROGLU, M. Chemical determination of free radical-induced damage to DNA. **Free Radic. Biol. Med**. v. 10, p. 225-242; 1991.

DO VALLE GOTTLIEB et. al. Longevity and aging in Rio Grande do Sul state: a hysterical, ethnic and morbi-mortality profile. **Rev Bras Ger e Geront.** (UnATI. Impresso), v. 14, p. 15-28, 2011.

DONG, X.C. et al. Inactivation of hepatic Foxo1 by insulin signaling is required for adaptive nutrient homeostasis and endocrine growth regulation. **Cell Metab.**,v. 8, p.65-

76, 2008.

DONG, Y. et al. SIRT1 is associated with a decrease in acute insulin secretion and a sex specific increase in risk for type 2 diabetes in Pima Indians. **Mol Genet Metab.**, v. 104, n. 4, p. 661-665, dec. 2011.

DUTTAetafPredictors of extraordinary survival inthe Iowa established populations f or epidemiologic study of the elderly:cohort follow-up to "extinction". **J Am Geriatr Soc**. v. 59, n.6, p. 963-71, jun. 2011.2011.

EL KHOURY, D. et al. Beta-glucan: Health benefits in obesity and metabolic syndrome. **J. Nutr. Metab.,** art. no. 851362, 2012.

ESTERBAUER, H.; SCHAUR, R.J.; ZOLLNER, H. Chemistry and biochemistry of 4-hydroxynonenal, malonaldehyde and related aldehydes. **Free Radic. Biol. Med.**, v.11, p.81-128, 1991.

FAIRBAIRN, D.W.; OLIVE, P.L.; O'NEILL, K.L. The Comet Assay: A comprehensive review. **Mutat. Res**. v. 339, p. 37-59, 1995.

FALLIN, M.D.; MATTEINI, A. Genetic Epidemiology in Aging Research.**J Gerontol A Biol Sci Med Sci.**, v.64A, n.1, p.47 - 60, 2009.

FENECH, M. et al. Molecular mechanisms of micronucleus, nucleoplasmic bridge and nuclear bud formation in mammalian and human cells. **Mutagenesis**, v. 26, n. 1, p.125-32. jan. 2011.

FERGUSON, L.R. Antimutagens as cancer chemopreventive agents in the diet. **Mutat Res**., v.307, p. 395-410, 1994.

FIGARSKA, S.M.; VONK, J.M.; BOEZEN, H.M. SIRT1 polymorphism, long-term survival and glucose tolerance in the general population. **PLoS One**, v. 8, n.3, e.58636, 2013.

FLACHSBART, F. et al. Sirtuin 1 (SIRT1) sequence variation is not associated with exceptional human longevity. **Experimental Gerontology**, v. 41, p. 98-102, 2006.

FONTANA L.; KLEIN, S. Aging, Adiposity, and Calorie Restriction. **JAMA,** v. 297, p.986-994, 2007.

FORTNEY, K. et al. Genome-Wide Scan Informed by Age Related Disease Identifies Loci for Exceptional Human Longevity. **PLoS Genet.** v. 11, n. 12, dec. 2015.

FRANCESCHI, C. et al. Inflammaging and anti-inflammaging: a systemic perspective on aging and longevity emerged from studies in humans, **Mech. Ageing Dev**., v.128, p. 92-105, 2007.

FRANZKE, B. et al. The impact of six months strength training, nutritional supplementation or cognitive training on DNA damage in institutionalized elderly. **Mutagenesis,** v. 30, p.147-153, 2015.

FREITAS, F. et al. Characterization of an extracellular polysaccharide produced by a Pseudomonas strain grown on glycerol. **Bioresour Technol**., v. 100, n. 2, p. 859-65. jan. 2009.

FRYE, R.A. Phylogenetic classification of prokaryotic and eukaryotic Sir2-like proteins. **Biochem Biophys Res Commun.**, v. 273, n. 2,p. 793-798, 2000.

FUKUSHIMA, M. et al. Hepatic LDL receptor mRNA in rats is increased by dietary mushroom (*Agaricus bisporus*) fiber and sugar beet fiber. **J. Nutr**., v. 130, n. 9, p. 2151-2156, 2000.

FUZIKAWA, A.K. et al.Association of ApoE polymorphisms with prevalent hypertension in 1406 older adults: the Bambui Health Aging Study (BHAS). **Brazil J Med and Biol Res**, v. *41*, n. 2, p. 89-94, 2008.

GENTSCHEW, L. et al. Polymorphisms in the superoxidase dismutase genes reveal no association with human longevity in Germans: a case-control association study.

Biogerontology, v. 14,n.6, p.719-727, dec. 2013.

GIACOMIN, K.C. et al. Bambui Project: a population-based study of the prevalence and factors associated with the need for a caregiver among the elderly. **Cad.**

Saùde Pùblica, Rio de Janeiro, v. 21, n. 1, p. 80-91, feb. 2005.

GIBSON, G.R. et al. Dietary modulation of the human colonic microbiota:updating the concept of prebiotics. **Nutr Res Rev**, v. 17, p. 259-275, 2004.

GIESE, E.C. et al. Free-radical scavenging properties and antioxidant activities of botryosphaeran and some other β-D-glucans. **Int J Biol Macromol**. v. 72, p.125130., Jan. 2015.

GOLUBEV, A.G. The other side of metabolism: a review. **Biochemistry**, v.61, p.1443-1460, 1996.

GOTTLIEB, M.G.V.; BONARDI, G.; MORIGUCHI, E.H.Physiopathology and inflammatory aspects of atherosclerosis. **Sci Med.,** Porto Alegre: PUCRS, v. 15, n. 3, p. 203207, jul./set. 2005.

GUTIÉRREZ, A.; PIETRO, A.; MARTiNEZ, A. T. Structural characterization of extracellular polysaccharides produced by fungi from the genus *Pleurotus*.

Carbohydr Res., v. 281, p. 143-154, 1996.

HACKAM, D.G ; ANAND, S.S. Emerging risk factors for atherosclerotic vascular disease: a critical review of the evidence. **JAMA**, v. 290, n. 7, p. 932-940, aug. 2003.

HALL, J.A. et al.The sirtuin family's role in aging and age-associated

pathologies. **The Journal of Clinical Investigation**, v. 123, n.3, p. 973-979, 2013.

HALLIWELL, B. Biochemical mechanisms accounting for the toxic action of oxygen on living organisms: the key role ofsuperoxide dismutase. **Cell Biol Int Rep**, v. 2, n. 2, p.113-28, mar.1978.

HALPERT, B.P. Development of the term "senility" as a medical diagnosis. **Minn Med**, v. 66, p. 421-424, 1983.

HAN, J. et al. Genetic variation in Sirtuin 1 (SIRTl)

is associated with lipid profiles but not with longevity in Ashkenazi Jews. **Transl Res.**,v.165, n. 4, p.480-1, apr. 2015.

HANSSON, G.K. Inflammation, atherosclerosis, and coronary artery disease. **N Engl J Med**, v. 352, n.16, p. 1685-1695, apr. 2005.

HARMAN, D. Aging: a theory based on free radical radiation chemistry. **J Gerontol**,

v. 11, p. 298-300, 1956.

HARMAN, D. The aging process: major risk factor for disease and death. **Proc. Natl. Acad. Sci. U.S.A.**, v.88, p. 5360-5363, 1991.

HAVEMAN-NIES, A.; DE GROOT, L.C.; VAN STRAVEREN, W.A. Dietary quality, lifestyle factors and healthy ageing in Europe: The SENECA study. **Ageing,** v. 32, p. 427-434, 2003.

HAYASHI, M. et al. The micronucleus assay with mouse peripheral blood reticulocytes using acridine orange-coated slides. **Mutat. Res**., v. 245, n.4, p. 245249, dec. 1990.

HEDDLE, J. A. et al. The induction of micronuclei as a measure of genotoxicity. A report of the U.S. Environmental Protection Agency Gene-Tox Program. **Mutat Res**., v. 123, n.1, p. 61-118, sep.1983.

HERRANZ, D. et al.

Sirt1 improves healthy aging and protects from metabolic syndrome- associated cancer. **Nat Commun**., p.1-3, apr. 2010.

HERSKIND, A.M. et al. The heritability of human longevity: a population-based study of 2872 Danish twin pairs born 1870-1900. **Hum Genet**.,v. 97, n.3,p. 319-23, mar. 1996.

HINERFELD, D. et al.Endogenous mitochondrial oxidative stress: neurodegeneration, proteomic analysis, specific respiratory chain defects, and efficacious antioxidant therapy in superoxide dismutase 2 null mice. **J Neurochem**. v. 88, p. 657-667, 2004.

HJELMBORG, JvB. J. et al. Genetic influence on human lifespan and longevity.**Hum Genet**., v.119, n.3, p. 312-21, apr. 2006.

HODGE, A.M. et al. Dietary patterns as predictors of successful ageing. **J Nutr Health Aging**, v. 18, p. 221-227, 2014.

HOLLEY, A.K. et al. Manganese superoxide dismutase: beyond life and death. **Amino acids**, v. 42, n.1,p.139-158, 2012.

HONG, L.; XUN, M.; WUTONG, W. Anti-diabetic effect of a beta-glucan from fruit body of maitake (*Grifola frondosa*) on KK-Ay mice. **J. Pharm. Pharmacol.**,v. 59, n.4, p. 575-582, 2007.

HOOG, N. et al. Inhibition of low-density lipoprotein oxidation by nitric oxide.

FEBS Lett, v.334, p.170-174, 1993.

HOUTHOOFD, K. et al. Extending life-span in *C. elegans*. **Science,** v. 305, p.12381239, 2004.

HUMPHREYS, V. et al. Age-related increases in DNA repair and antioxidant protection: a comparison of the Boyd Orr Cohort of elderly subjects with a younger population sample, **Ageing**, v. 36, p.521-526, 2007.

HUANG, R.Y. et al. Hydrogen/deuterium exchange and electron-transfer dissociation mass spectrometry determine the interface and dynamics of apolipoprotein E oligomerization. **Biochemistry**, v. 50, p. 9273-9282, 2011.

HYLAND, P. et al. Nonagenarians from the Swedish NONA Immune Study have increased plasma antioxidant capacity and similar levels of DNA damage in peripheral blood mononuclear cells compared to younger control subjects. **Exp. Gerontol.**, v. 37, 465-473, 2002.

IBGE. Brazilian Institute of Geography and Statistics. Projection of the population of Brazil, by sex and age - 1980-2050. Rio de Janeiro, Brazil, 2008. 93p.

IBGE. Brazilian Institute of Geography and Statistics. Synthesis of Social Indicators. An analysis of the living conditions of the Brazilian population. Rio de Janeiro, Brazil, 2013a. 266p

IBGE. Brazilian Institute of Geography and Statistics. Methodological Reports Series. v.40. Population Projections - Brazil and Federative Units. Rio de Janeiro, Brazil, 2013b. 41p.

IBGE. Brazilian Institute of Geography and Statistics. Studies and Analyses - Demographic and Socioeconomic Information number 3. Demographic Change in

Brazil at the Beginning of the 21st Century. Subsidies for population projections. Rio de Janeiro, Brazil, 2015. 156p.

ISHIBASHI, S. et al. Hypercholesterolemia in low density lipoprotein receptor knockout mice and its reversal by adenovirus-mediated gene delivery. **J Clin Investv.** 92, p. 883-893, 1993.

JACOB FILHO, W. Physical activity and healthy aging. Revista Brasileira de Educaçao Fisica Especializada. Sao Paulo, v.20, p.73-77, supplement n. 5, set. 2006.

JACOBSEN, R. et al. Increased effect of the ApoE gene on survival atadvanced age in healthy and long-lived Danes: two nationwide cohort studies.**Aging Cell**, v. 9, p.1004-1009, 2010.

JANG, S. et al. Scavenger receptor CL-P1 mediates endocytosis by associating with AP-2μ2. **Biochim Biophys Acta**, n. 1840, supp.11, p. 3226-3237, nov. 2014.

JAWIEN, J.; NASTALEK, P.; KORBUT, R. Mouse Models of Experimental Atherosclerosis. **J Physiol Pharmacol.**, v. 55, n.3, p. 503-517, sept. 2004.

JEMAI, H. et al. Lipid-lowering and antioxidant effects of hydroxytyrosol and its triacetylated derivative recovered from olive tree leaves in cholesterol-fed rats. **J Agric Food Chem.**, v. 23, n. 56, supp. 8, p. 2630-2636, apr. 2008.

JOHNSON, T.E. *et al.* Relationship between increased longevity and stress resistance as assessed through gerontogene mutations in *Caenorhabditis elegans*. **Exp. Gerontol.** v. 36, p.1609-1617, 2001.

JOHNSON, T. Genes, phenes, and dreams of immortality: the 2003 Kleemeier Award Lecture. **J Gerontol**, v. 60A, n. 6, p.680-687, 2005.

KADA, T., MORITA, K., INOUE, T. Anti-mutagenic action of vegetable factor(s) on the mutagenic principle of tryptophan pyrolysate. **Mutat Res.**, Amsterdam, v.53, n.3, p.351-353, 1978.

KADAM, S.S.; MOHADIK, K.R.; BOTHARA, K.G. Anti-neoplastic agents. In: LEMKE, T.L. et al (Ed.) **Foye's Principles of med chem.** 6. ed. Bangalori: Nerali

Prakashan, v. 1, 2007. p. 141-166.

KAGIMURA, F.Y. et al. Biological activities of derivatized D-glucans: a review. **Int J Biol Macromol.** v.72, p. 588-98, Jan. 2015.

KAHLON, T.S.; WOODRUFF, C.L. In vitro Binding of Bile Acids by Rice Bran, Oat Bran, Barley and β-Glucan Enriched Barley. **Cereal Chem.**,v. 80, n.3, p. 260263, 2003.

KAMINSKI, M.M. et al. Mitochondria as oxidative signaling organelles in T-cell activation: physiological role and pathological implications. **Arch Immunol Ther Exp**. (Warsz), v.61, p. 367-384, 2013.

KAPAHI P.; ZID, B. TOR pathway: linking nutrient sensing to life span. **Sci Aging Knowledge Environ.**, v. 36, PE34, sept. 2004.

KATO, D. et al. Antiviral activity of chondroitin sulphate E targeting dengue virus envelope protein**. Antiviral Res**, v. 88, p. 236-243, 2010.

KAUR, N.; GUPTA, A.K. Applications of inulin and oligofructose in health and nutrition. **J. Biosci**., Bangalore, v. 27, p. 703-714, 2002.

KAZIMIROVA, A. et al. Does a vegetarian diet influence genomic stability? **Eur J Nutr**., v. 43, p.32-38, 2004.

KERBER, R.A. et al. Familial excess longevity in Utah genealogies. **J Gerontol A Biol Sci.** v. 56, n. 3, p.B130-B139, 2001.

KERDILES, YM, et al. Foxo transcription factors control regulatory T cell development and function. **Immunity**,v. 33, p. 890-904, 2010.

KILIC, U. et al. A Remarkable Age-Related Increase in SIRT1 Protein Expression against Oxidative Stress in Elderly: *SIRT1* Gene Variants and Longevity in Human. **Plos One**, v. *10*, n. 3, p.1-19, 2015.

KING, C.M. et al. In vivo antioxidant status, DNA damage, mutation and DNA repair capacity in cultured lymphocytes from healthy 75- to 80-year-old humans, **Mutat. Res**. v. 377, p.137-147, 1997.

KINSELLA, K.; HE, W. An Aging World: 2008. International Population Reports. NATIONAL INSTITUTE ON AGING.U.S. issued: June, 2009.

KIM, D.H. et al. Apreliminary study on the hypoglycemic effect of the exopolysaccharides produced by five different medicinal mushrooms. **J Microbiol Biotechnol.,** v. 11, p. 167-171, 2001.

KIM, Y.W. et al. Anti-diabetic activity of beta-glucans and

their enzymatically hydrolyzed oligosaccharides from *Agaricus blazei*. **Biotechnol Lett.**, v. 27, n.7, p. 483-487, apr. 2005.

KIRSCH-VOLDERS, M.; VANHAUWAERT, A.; DE BOECK, M.; DECORDIER,

I. Importance of detecting numerical versus structural chromosome aberrations.

Review. **Mutat Res**. v. 25, n. 504, p. 137-148, jul. 2002.

KLAss, M.R. A method for the isolation of longevity mutants in the nematode Caenorhabditis elegans and initial results. **Mech. Ageing Dev**. 22, 279-286,1983.

KOBAYAsHI, H. et al. A comparison between manual microscopic analysis and computerized image analysis in the single cell gel electrophoresis assay. **MMS Commun**. , v. 3, p.103-115, 1995.

KOHEN, R.; NYSKA, V. Oxidation of biological systems: oxidative stress phenomena, antioxidants, redox reactions, methods for their quantification. **Toxicol Pathol**, v. 30, n. 6, p. 620-650, 2002.

KOROU, L.M. etal. Comparative antilipidemic effect of N-acetylcysteine and sesaem oil administration in diet-induced hypercholesterolemic mice. **Lipids Health Dis.**, v. 9, p.23, 2010.

KOWALTOWSKI, A.J.; CASTILHO, R.F.; VERCESI, A.E. Mitochondrial permeability transition and oxidative stress. **FEBS Lett**., v. 495, p. 12-15, 2001.

KRCMAR, P. et al. Structure of extracellular polysaccharide produced by lignindegrading fungus Phlebia radiata in liquid culture. **Int J Biol Macromol.,** v.24, n.1, p. 61-64, jan. 1999.

KRISHNA, G.; HAYASHI, M. *In vivo* rodent micronucleus assay: protocol, conduct and data interpretation. **Mutat. Res.**, n. 455, p. 155-166, 2000.

KUNINGAS, M. et al. SIRT1 Gene, Age-Related Diseases, and Mortality: The Leiden 85-Plus Study. **Journal of Gerontology: biological sciences**, v. 62A, n.. 9, p.960-965, 2007.

LABAER, J. So, you want to look for biomarkers (introduction to the special biomarkers issue). **J Proteome Res**., v. 4, n.4, p.1053-1059, jul.-aug. 2005.

LABAT-ROBERT, J. ; ROBERT, L. Longevity and aging. Role of free radicals and xanthine oxidase. A review. **Pathol Biol** (Paris), v.62, n.2, p.61-66, apr. 2014.

LAM, E.W.; FRANCIS, R.E.; PETKOVIC, M. FOXO transcription factors: key regulators of cell fate. **Biochem Soc Trans**., v. 34, pt. 5, p.722-726, nov. 2006.

LAM, Y.Y.; PETERSON, C.M.; RAVUSSIN, E. Resveratrol vs. calorie restriction: data from rodents to humans. **Exp Gerontol.**, v. 48, n.10, p.1018-24, oct. 2013.

LAZARIDOU, A.; BILIADERIS, C.G. Molecular aspects of cereal beta-glucan functionality: Physical properties, technological applications and physiological effects. **J. Cereal Sci.,** v. 46, p. 101-118, 2007.

LAZARIDOU, A. et al. A micro- and macro-scale approach to probe the dynamics of sol-gel transition in cereal beta-glucan solutions varying in molecular characteristics. **Food Hydrocol**., v. 39,p. 204-214, 2014.

LEBRAO, M.L.; LAURENTI, R. Saúde, bem-estar e envelhecimento: o estudo SABE no Municipio de Sao Paulo. **Rev. Bras. Epidemiol**., Sao Paulo, v. 8, n. 2, p.127-141, jun. 2005.

LEE, S.S. et al. DAF-16 target genes that control C. elegans life-span and metabolism. **Science**, v. 300, p.644-647, 2003.

LEON, L.R. The use of gene knockout mice in thermoregulation studies. **J Therm Biol.**, v. 30, p. 273-88, 2005.

LIBBY, P. Inflammation in atherosclerosis. **Nature**, v. 420, n. 6917, p. 868-874, dec.

2000.

LIMA, T.A.S.; MENEZES, T.M.O. Investigating the production of knowledge about the long-lived elderly. **Rev Brâs Enferm.**, v. 64, n.4, p. 751-758, Jul.-Aug. 2011.

LIMA E COSTA, M.F. et al.The Bambui health and ageing study (BHAS): methodological approach and preliminary results of a population-based cohort study of the elderly in Brazil. **Rev Saù Pùbl**, v.34, n.2, p.126-135, 2000.

LIMA-COSTA, M.F.; VERAS, R. Saúde pùblica e envelhecimento. **Cad Saùde Pùblica**, v.19, p.700-701, 2003.

LIN, S.; DEFOSSEZ, P.; GUARENTE, L. Requirement of NAD and SIR2 for lifespan extension by calorie restriction in *Saccharomyces cerevisiae*. **Science**, v.289, p.2126-2128, 2000.

LIU, Y. et al. Antihyperglycemic, antihyperlipidemic and antioxidant activities of polysaccharides from *Catathelasma ventricosum* in streptozotocin-induced diabetic mice. **Food Chem. Toxicol.**, v.57, p. 39-45, 2013.

LOMBARD, D.B. et al. DNA repair, genome stability, and aging**. Cell**, v. 120, p. 497-512, 2005.

LONDON, L.E. et al. Exopolysaccharide-producing probiotic Lactobacilli reduce serum cholesterol and modify enteric microbiota in ApoE-deficient mice. **J Nutr**, v.144, n. 12, p. 1956-1962, dec. 2014.

LUCAS, E.H. et al. Production of oncostatic principles in vivo and in vitro by species of the genus Calvatia. **Antibiot Annu**, v. 6, p. 493-496, 1958.

LUSIS, A.J. Atherosclerosis. **Nature,** v. 14, n. 407 (6801), p. 233-241, sep. 2000.

LUZHNA, L.; KATHIRIA, P.; KOVALCHUK, O. Micronuclei in genotoxicity assessment: from genetics to epigenetics and beyond. **Front Genet.,** v. 4, n.131, p.117, jul. 2013 Jul.

MACGREGOR, J.T.; HEDDLE, J.A.; HITE, M.; MARGOLIN, B.H.; RAMEL, C.; SALAMONE, M.F.; TICE, R.R.; WILD, D. Guidelines for the conduct of micronucleus assays in mammalian bone marrow erythrocytes. **Mutat Res.**, v.189, n. 2, p.103-121, oct. 1987.

MAKI et al. Lipid responses to consumption of a beta-glucan containing ready-to-eat cereal in children and adolescents with mild-to-moderate primary hypercholesterolemia. **Nut. Res.,** v. 23, p. 1527-1535, 2003.

MAHLEY, R.W. Apolipoprotein E: cholesterol transport protein with expanding role in cell biology. **Science**, v. 240, p. 622-30, 1988.

MAHLEY, R.W.; HUANG, Y.; RALL, S.C. Pathogenesis of type III hyperlipoproteinemia (dysbetalipoproteinemia): questions, quandaries, and paradoxes. **J Lipid Res**, v. 40, p.1933-1949, 1999.

MAHLEY, R.W.; RALL JR, S.C. Apolipoprotein E: far more than a lipid transport protein.

Annual review of genomics and human genetics, v. 1, p. 507-537, 2000.

MALINI, M. et al. Modulation of gene expression and cell cycle by botryosphaeran, a (1→3)(1→6)-β-d-glucan in human lymphocytes. **Int J Biol Macromol**, v. 77, p. 214-221, 2015.

MANTOVANI, M.S. et al. β-Glucans in promoting health: Prevention against mutation and cancer. **Mutat Res.**, v. 658, n.3, p. 154-161, 2008.

MARNETT, L.J. Lipid peroxidation-DNAdamageby malondialdehyde.

Mutat.Res.,v. 424, p.83-95, 1999.

MARTINICHEN-HERRERO, J.C. Et al. Anticoagulant and antithrombotic activities of a chemically sulfated galactoglucomannan obtained from the lichen Cladonia ibitipocae. **Int J Biol Macromol**. v. 35, n. 1-2, p. 97-102, mar. 2005.

MARX, J. Nobel Prize in Physiology or Medicine. Tiny worm takes a star tune. **Science**, v. 288, n. 5593, p. 526, oct. 2002.

MATÉS, J.M.; PÉREZ-GOMES, C.; CASTRO, I.N. Antioxidant enzymes and human diseases. **Clin Biochem.**, v.32, p.595-605, 1999.

MCCAULEY, B.S.; DANG, W. Histone methylation and aging: Lessons learned from model systems. **Biochim et Biophys Acta**, v. *1839*, n. 12, p.1454-1462, 2014.

MCGUE, M.et al. Longevity is moderately heritable in a sample of Danish twins born 1870-1880. **J Gerontol,** v.48, n.6, p. B237-B244, 1993.

MCINTOSH, M.; STONE, B.A.; STANISICH, V.A. Curdlan and other bacterial (1→3)-beta-D-glucans. **Appl Microbiol Biotechnol.**, v. 68, p.163-173, 2005.

MECOCCI, P. et al. Plasma antioxidants and longevity: a study on healthy centenarians. **Free Radic Biol Med.**, v. 28,p.1243-1248, 2000.

MERCADO-SAENZ, S. et al. Cellular aging: theories and technological influence. **Braz Arch Biol Technol**,, v. 53, n. 6, p.1319-1332, 2010.

MICHAN, S.; SINCLAIR, D. Sirtuins in mammals: insights into their biological function. **Biochem J.**, v. 404, p.1-13, 2007.

MIRANDA, C.C. et al. Anticlastogenic activity exhibited by botryosphaeran, a new exopolysaccharide produced by Botryosphaeria rhodina MAMB-05. **Int J Biol Macromol.**, v. 1, n. 42, sup. 2, p.172-177, mar. 2008.

MIRANDA-NANTES, C.C. et al. Hypoglycemic and hypocholesterolemic Effects of Botryospahaeran form Botryosphaeria MAMB-05 in Diabetes-Induced and Hyoerlipidemia conditions in Rats. **Mycobiology**, v. 39, n.3, p. 187-193, sep. 2011.

MITCHELL, B.D. et al. Heritability of life span in the Old Order Amish. **Am J Med Genet**, v. 102, n. 4, p. 346-352, 2001.

MOREAU, R. et al. Reversal by aminoguanidine of the age-related increase in glycoxidation and lipoxidation in the cardiovascular system of Fischer 344 rats.

Biochem.Pharmacol., v.69, p.29-40, 2005.

MORI, K. et al. Scavenger receptor CL-P1 mainly utilizes a collagen-like domain to uptake microbes and modified LDL**. Biochim Biophys Acta**, n. 1840, supp. 12, p. 3345-56, dec. 2014.

MURPHY, C.T. et al. Genes that act downstream of DAF-16 to influence the lifespan of *Caenorhabditis elegans*. **Nature**, v. 424, p.277-283, 2003.

MUTLU-TURKOGLU, U. et al. Age-related increases in plasma malondialdehyde and protein carbonyl levels and lymphocyte DNA damage in elderly subjects. **Clin Biochem.**,v. 36, p. 397-400, 2003.

MURABITO, J.M.; YUAN, R.; LUNETTA, K.L. The Search for Longevity and Healthy Aging Genes: Insights From Epidemiological Studies and Samples of Long-Lived Individuals. **J Gerontol A Biol Sci Med Sci**, v. 67A, n. 5, p.470-479, may 2012.

NAPOLI, C.; IGNARRO, L.J. Nitric oxide and pathogenic mechanisms involved in the development of vascular diseases. **Arch Pharm Res, v.** 32,n. 8,p.1103-1108, 2009.

NEGRE-SALVAYRE, A. et al. Advanced lipid peroxidation end products in oxidative damage to proteins. Potential role in diseases and therapeutic prospects for the inhibitors. **Br. J. Pharmacol.**, v.153, p. 6-20, 2008.

NERI, M. et al. Worldwide interest in the comet assay: a bibliometric study. **Mutagenesis**, v. 30, p. 155-163, 2015.

NEWMAN, A.B.; MURABITO, J.M. The epidemiology of longevity and exceptional survival. **Epidemiol. Rev.**, v. 35, p.181-197, 2013.

NIELSEN, F .et al. Plasma malondialdehyde as biomarker for oxidative stress: reference interval and effects of life-style factors. **Clin Chem.**,v.43, n.7, p.1209-14, jul.1997.

NOGUEIRA, S.L. et al. Spatial distribution and growth of the elderly population in Brazilian capitals 1980 to 2006: an ecological study. **Rev. Bras. Estud. Popul.**, v.25, n.1, p.195-198, 2008.

NOGUEIRAS, R., et al. Sirtuin 1 and Sirtuin 3: Physiological Modulators Of Metabolism. **Physiol Rev.,** v. 92, n.3, p.1479-1514, 2012.

NOVAK, M.; VETVICKA,V. Beta-glucans, history, and the present: immunomodulatory

aspects and mechanisms of action. **J Immunotoxicol.,** v.5, n.1, p. 47-57, jan. 2008.

NOVAK, M.; VETVICKA, V. Glucans as biological response modifiers. **Endocr Metab Immune Disord Drug Targets**, v. 9, n. 1, p. 67-75, mar. 2009.

NOZIK-GRAYCK, E.; SULIMAN, H.B.; PIANTADOSI, C.A.

Extracellular superoxide dismutase. **Int J Biochem Cell Biol.**, v. 37, n.12,p. 24662471, dec. 2005.

O'BRIEN T.; NGUYEN, T.T. Lipids and lipoproteins in women. **Mayo Clin Proc**, v. 72, p. 235-244, 1997.

OECD. Organization for Economic Cooperation and Development Guideline for Genetic Toxicology. Testing of Chemicals: Mammalian erythrocyte micronucleus test. Paris, Guideline 474, 1997. 10 p.

OLIVEIRA, H.C. et al. Oxidative stress in atherosclerosis prone mouse is due to low antioxidant capacity of mitochondria. **FASEB J.**, v. 19, p. 278-280, 2005.

WHO. World Health Organization. Active ageing: a health policy. Translated by Suzana Gontijo. Brasilia: Pan American Health Organization, 2005. p.33-44.

WHO. World Health Organization. A glossary of terms for community health care and services for older persons. 2004. Ageing and Health Technical Report, volume 5. p. 42. Available at: <http://www.who.int/>. Accessed on January 20, 2015.

WHO. World Health Organization. Mental health and older adults. Fact sheet N°381. september 2015. Available at: <http://www.who.int/>. Accessed on: 15 Jan 2015.

OSTLING, O.; JOHANSON, K. J. Microelectrophoretic study of radiation-induced DNA damage in individual mammalian cells. **Biochem. Biophys. Res. Commun.**, v.123, p. 291-298, 1984.

PAMPLONA, R. et al. Proteins in human brain cortex are modified by oxidation, glycoxidation, and lipoxidation. Effects of Alzheimer's disease and identification of lipoxidat ion targets. **J. Biol.Chem**. v. 280, p.21522-21530, 2005.

PALLONI, A.; PINTO-AGUIRRE, O.; PELAEZ, M. Demographic and health conditions of ageing in Latin America and the Caribbean. **Int J Epidemiol**, v. 31, n. 4,

p. 762-71, 2002.

PAREDES SALIDO, F.; ROCA FERNANDEZ, J. J. Influence of free radicals on cellular aging. **Bioquimica**, v. 21, p. 96-100, 2002.

PEREIRA, D.I.A.; GIBSON, G.R. Effects of consumption of probiotics and prebiotics on serum lipid levels in humans. *Critical* **Rev. Biochem. Molec. Biol.**, v. 37, n. 4, p.259-281, 2002.

PERLS, T.T. et al. Siblings of centenarians live longer. **Lancet,** v. 23, n. 351(9115), p.1560, may 1998.

PETRAVIC-TOMINAC, V. et al. Biological Effects of Yeast β-Glucans. **Agric Conspec Sci.,** v. 75, n.4, p. 149-158, 2010.

PRATICÒ, DOMENICO. Lipid Peroxidation and the Aging Process. Review. **Science of Aging Knowledge Environment**, p. 1-4, 18, dec. 2002.

POULAIN, M.; PES, G.; SALARIS, L. A Population Where Men Live As Long As Women: Villagrande Strisaili, Sardinia. **J Aging Res.**, article ID 153756, 10 p., 2011.

QUEIROZ, E.A. et al. Antiproliferative and pro-apoptotic effects of three fungal exocellular β-glucans in MCF-7 breast cancer cells is mediated by oxidative stress,

AMP-activated protein kinase (AMPK) and the Forkhead transcription factor, FOXO3a. **Int J Biochem Cell Biol.**, v. 67, p. 14-24, oct. 2015.

RAA, J. Immune modulation by non-digestible and non-absorbable beta-1,3/1,6-glucan.

Microb Ecol Health Dis., v. 29, n. 26: 27824, may 2015.

RAJPATHAK, S.N. et. al. Lifestyle Factors of People With Exceptional Longevity. **J Am Geriatr Soc.,** v. 59, n.8, p. 1509-1512, 2011.

RAMESH, H.P.; THARANATHAN, R.N. Carbohydrates the renewable raw materials of high biotechnological value. **Crit Rev Biotechnol.**, v. 23, n.2, p.149-73, 2003.

RIBARIC, S. Diet and Aging. Review. **Oxid Med Cell Longev**., v. 2012, 741468, p.1-20, 2012.

RIBEIRO, S.M.L.; CAMPOS, P.; TIRAPEGUI, J. The rat as a laboratory animal: history, biological data and critical analysis of its use. **Rev Farm Bioquim Univ Sao Paulo**, v. 31, n. 1, p. 21-28, 1995.

RICHARDSON, A. et al. Measures of Healthspan as Indices of Aging in MiceA Recommendation. **J Gerontol A Biol Sci Med Sci**, v. 71, n. 4, p. 427-430, apr. 2016.

ROEHRICH, M.E. et al. Insulin-secreting beta-cell dysfunction induced by human lipoproteins. **J Biol Chem**, v. 278, p.18368-18375, 2003.

ROSADO, F.R. et al. Biomass and exopolysaccharide production in submerged cultures of *Pleurotus ostreatoroseus* SING., and *Pleurotus ostreatus* "florida" (JACK.: FR.) KUMMER. **J. Basic Microbiol.**, v. 43, p. 230-237, 2003.

ROSS, R. Atherosclerosis-an inflammatory disease. **N Engl J Med**, v. 340, n.2, p. 115-126, jan.1999.

ROSS, G. D., et al. Therapeutic intervention with complement and β-glucan in cancer.

Immunopharmacology, v. 42, p. 61-74, 1999.

RUBEL, R. et al. Hypolipidemic and antioxidant properties of *Ganoderma lucidum* (Leyss:Fr) Karst used as a dietary supplement. **World J. Microbiol. Biotechnol.**, v. 27, n. 5, p. 1083-1089, 2011.

RUSSO, P. et al. Beta-Glucans Improve Growth, Viability and Colonization of Probiotic Microorganisms. **Int. J. Mol. Sci.**, v. 13, p.6026-6039, 2012.

RÜTTI, S. et al. Low- and high-density lipoproteins modulate function, apoptosis, and proliferation of primary human and murine pancreatic betacells. **Endocrinology**, v. 150, p. 4521-4530, 2009.

SAITO, H. et al. Effects of polymorphism on the lipid interaction of human apolipoprotein E. **J Biol Chem.**, v. 278, p. 40723-40729, 2003.

SANTOS, B.F. Animal Model. In: ANDRADE, A.; PINTO, S.C.; OLIVEIRA, R.S.

Laboratory Animals: breeding and experimentation. Rio de Janeiro: Editora Fiocruz, 2002a. chap. 2. p. 23-24.

SANTOS, B.F. Breeding and management of mice. In: ANDRADE, A.; PINTO, S.C.; OLIVEIRA, R.S. **Animais de Laboratório: criação e experimentação.** Rio de Janeiro: Editora Fiocruz, 2002b. chap. 14. p. 115-118.

SAWANT, S.G.; FIELDEN, M.R.; BLACK, K.A. Evaluation of genotoxicity testing of FDA approved large molecule therapeutics **Regul Toxicol Pharmacol.**, v. 70, n.1, p. 87-97, oct. 2014.

SCHLEGEL,R.;MACGREGOR,J.T.The persistence of micronuclei in peripheral blood erythrocytes: detection of chronic chromosome breakage in mice. **Mutat Res**., v.104, n.6, p. 367-369, jul. 1982.

SCHMID, W. The micronucleus test. **Mutat Res**., v.31, n.1, p.9-15, feb. 1975.

SCHUMACHER, B.; GARINIS, G.A.; HOEIJMAKERS, J.H. Age to survive: DNA damage and aging. **Trends Genet.**, v. 24, p. 77-85, 2008.

SCHUPF, N. et al. Apolipoprotein E and familial longevity. **Neurobiol. Aging,** v. 34, p.1287-1291, 2012.

SCHWANKE, C.H.A. et al. Analysis of the association between apolipoprotein E gene polymorphism and cardiovascular risk factors in the long-lived elderly. **Arq Bras Cardiol**, v. 78, n.6, p. 561-579, jun. 2002.

SEARS, D.A.; UDDEN, M.M. Howell-Jolly bodies: a brief historical review.

Am.J.Med. Sci. 343, 407-409, 2011.

SEBASTIANI, P.; PERLS, T.T. The genetics of extreme longevity: lessons from the New England Centenarian Study. **Front Genet**., v. 3, p.277, 2012.

SEBASTIANI, P. et al. Meta-analysis of genetic variants associated with human exceptional longevity. **Ageing**, v. 5, p. 653-661, 2013.

SEDELNIKOVA, O.A. et al. DNA double-strand breaks form in bystander cells after microbeam irradiation of three-dimensional human tissue models. **Cancer Res.**, v. 67, n. 9, p. 4295-4302, may 2007.

SELBMANN, L. et al. Production and structural characterization of the exopolysaccharide of the Antarctic fungus Phoma herbarum CCFEE 5080. **Res Microbiol.**, v.153, p. 585-592, 2002.

SELBMANN, L.; STINGELE, F.; PETRUCCIOLI, M. Exopolysaccharide production by filamentous fungi: the example of Botryosphaeria rhodina. **Antonie van Leeuwenhoek**, v. 84, n. 2, p.135-45, 2003.

SHAH, V. et al. Characterization of the extracellular polysaccharide produced by a marine cyanobacterium, Cyanothece sp. ATCC 51142, and its exploitation toward metal removal from solutions. **Curr Microbiol.**, v.40, n.4, p. 274-278, apr. 2000.

SHIMODA-MATSUBAYASHI, S. et al. Structural Dimorphism in the Mitochondrial Targeting Sequence in the Human Manganese Superoxide Dismutase Gene. A Predictive Evidence for Conformational Change to Influence Mitochondrial Transport and a Study of Allelic Association in Parkinson's Disease. **Biochem.**

Biophys. Res Commun. v. 226, p. 561-5, 1996.

SILVA, M. de L. C. da et al. Chemical characterization of fungal glucans and their biotechnological applications. **Quim. Nova**, v. 29, n.1, p. 85-92, 2006.

SIMM, ANDREAS; KLOTZ, LARS-OLIVER. Stress and biological aging - A double-edged sword**. Z Gerontol Geriat,** v. 48, p. 505-510, 2015.

SIMMONS, D. Epigenetic Influences and Disease. **Nature Education**, v. 1, n. 1, p.6, 2008.

SINGH, N.P. et al. Abundant alkali

sensitive sites in DNA of human and mouse sperm. **Exp Cell Res.,** v. 184, n. 2, p. 461-70, oct. 1989.

SMITH , J.S. et al. A phylogenetically conserved NAD+-dependent protein deacetylase activity in the Sir2 protein family. **Proc Natl Acad Sci USA**, v. 97, p. 6658-6663, 2000.

SOARES, J.P. et al. Aging and DNA damage in humans: a meta-analysis study. **Aging** (Albany NY), v.6, n.6, p. 432-9, jun. 2014.

SOERENSEN, M .et al. The Mn superoxide dismutase single nucleotide polymorphism rs4880 and the glutathione peroxidase 1 single nucleotide polymorphism rs1050450 are

associated with aging and longevity in the oldest old. **Mech Ageing Dev.**, v. 130, n.5, p. 308-314, may 2009.

SOERENSEN, M .et al. Replication of an association of variation in the FOXO3 gene with human longevity using both case-control and longitudinal data. **Aging Cell**, v. 9, n.6, p. 1010-1017, dec. 2010.

SOERENSEN, M .et al. Evidence from case-control and longitudinal studies supports associations of genetic variation in *APOE*, *CETP*, and *IL6* with human longevity. ***Age***, v. *35*, n. 2, p.487-500, 2013.

SOERENSEN, M. et al.Association study of FOXO3 SNPs and aging phenotypes in Danish oldest-old individuals. **Aging Cell**, v. 14, n.1, p. 60-66, feb. 2015.

SOHAL, R.S.; WEINDRUCH, R. Oxidative stress, caloric restriction, and aging.

Science, v. 273, n. 5271, p. 59-63, 1996.

SOLTANIAN, S. et al. Beta-glucans as immunostimulant in vertebrates and invertebrates. **Crit Rev Microbiol**., v.35, n.2, p.109-38. 2009.

SPATAREANU, A. et al. Synthesis, characterization and solution behaviour of oxidized pullulan. **Carbohydr Polym**.v.13, n. 111, p. 63-71, oct. 2014.

STEGHENS, J.P. et al. Diaminonaphtalene, a new highly specific reagent for HPLC-UV measurement of total and free malondialdehyde in human plasma or serum. **Free Radic Biol Med**., v.1, n.2, p. 242-249, jul. 2001.

STEINBERG, D. et al. Beyond cholesterol modifications of low-density lipoprotein that increase its atherogenicity. **NEJM**., v. 320, p. 915-924, 1989.

STELUTI, R.M. et al. Comparison of Botryosphaeran production by

the ascomyceteous fungus Botryosphaeria sp., grown on differentcarbohydrate carbo n sources, and their partial structural features**. J Basic Microbiol.**, v. 44, n. 6, p.

480486, 2004.

STESSMAN, J. et al. Candidate genes associated with ageing and life expectancy in the Jerusalem longitudinal study. **Mech Ageing Dev**., v. 126, n.2, p. 333-339, feb. 2005.

SUN, J. et al. Induced overexpression of mitochondrial Mn-superoxide dismutase extends the life span of adult *Drosophila melanogaster*. **Genetics,** v. 161, n.2, p. 661-672, 2002.

SURENJAV, U. et al. Effects of molecular structure on antitumor activities of (1→3)-*β* -D-glucans from different *Lentinus Edodes*. **Carbohydr. Polym**., v. 63, p. 97-104, 2006.

SUTTON, A. et al. The manganese superoxide dismutase Ala16Val dimorphism modulates both mitochondrial import and mRNA stability. **Pharmacogenet. Genomics**, v. 15, p. 311-319, 2005.

SWINDELL, W.R. et al. Study Of Osteoporotic Fractures Research Group.Indicators of "healthy aging" in older women (65-69 years of age). A data mining approach based on prediction oflong-term survival. **BMC Geriatr**., v. 10, n.55, p. 1-24, aug. 2010.

SYNYTSYA, A.; NOVAK, M. Structural diversity of fungal glucans. Review. **Carbohydr Polym.**, v. 92, 792- 809, 2013.

TAUFER, M. et al. Is the Val16Ala manganese superoxide dismutase polymorphism associated with the aging process? **J Gerontol A Biol Sci Med Sci**, v. 60, n.4, p. 432-438, apr. 2005.

TERRY, D.F. et al. Cardiovascular risk factors predictive for survival and morbidity-free survival in the oldest-old Framingham Heart Study participants. **J Am Geriatr Soc.**, v. 53, n.11, p. 1944-1950, 2005.

TIAN, C. ; FANG, S. ; DU, X. ; JIA, C. Association of the C47T polymorphism in SOD2 with diabetes mellitus and diabetic microvascular complications: a meta analysis. **Diabetologia**, v. 54, n. 4, p. 803-811, apr. 2011.

TICE, R.R. et al. Single cell gel/comet assay: guidelines for in vitro and in vivo genetic toxicology testing. **Environ Mol Mutagen**, v. 35, n.3, p. 206-221, 2000.

TINDALE, L.C. et al.. Rare and common variants in the Apolipoprotein E gene in healthy old age. **Neurobiol Aging,** v. 35, n.3, p.727.e1-e3, mar. 2014.

TRAN, H. et al. DNA repair pathway stimulated by the forkhead transcription factor FOXO3 through the Gadd45 protein. **Science**, v. 296, n. 5567, p. 530-534, 2002.

TWEATS, D.J.; et al. Report of the IWGT working group on strategies and interpretation of regulatory *in vivo* tests I. Increases in micronucleated bone marrow cells in rodents that do not indicate genotoxic hazards. **Mutat. Res.**, v. 627, p. 78-91, 2007.

TZIVION, G.; DOBSON, M.; RAMAKRISHNAN, G. FoxO transcription factors; Regulation by AKT and 14-3-3 proteins. **Biochim Biophys Acta**, v. 1813, n. 11, p. 1938-1945, nov. 2011.

UCHIDA, K. Role of reactive aldehyde in cardiovascular diseases. **Free Radic. Biol. Med.**, v. 28, p.1685-1696, 2000.

US FDA. US Food & Drug Administration. Food Labeling: health claims; soluble dietary fiber from certain foods and coronary heart disease. Federal Register, 67, 61773e61783,1997.

VALADARES, M.O.; VIANNA, L.G.; MORAES, C.F. The theme of human aging in Brazilian research groups. **Revista Kairós Gerontologia**, v.16, n. 2, p. 117-128, mar. 2013.

VANDERLAAN, A.P.; REARDON, A.C.; GETZ, S.G. Site specificity of atherosclerosis. **Arterioscler Thromb Vasc Biol**., v. 24, p.12-22, jan. 2004.

VASCONCELOS, A. F. et al. Sulfonation and anticoagulant activity of fungal exocellular β-(1→6)-D-glucan (lasiodiplodan). **Carbohydr Polym.**, v.15, n. 92, sup. 2, p.1908-1914, feb. 2013.

VASQUEZ, E.C. et al. Cardiac and vascular phenotypes in the apolipoprotein

Edeficient mouse. **J Biomed Sci.**, v. 13, p.19-22, 2012.

VETVICKA, V.; VETVICKOVA, Physiological effects of different types of beta glucan. **J. Biomed Pap Med Fac Univ Palacky Olomouc Czech Repub**., v. 151, n.2, p. 225-231, dec. 2007.

VON BORSTEL, R.C.; DRAKE, J.W.; LOEB, L.A. Foreword. **Mutat Res/Fund Mol Mech Mutag**., v. 350, n.1, p.1-3, 1996.

WANG, J. et al. Sulfated modification, characterization and structure-antioxidant relationships of Artemisia sphaerocephala polysaccharides. **Carbohydr Polym**, v. 81, p. 897-905, 2010.

WANG, Y. et al. High-molecular-weight β-glucan decreases serum cholesterol differentially based on the CYP7A1 rs3808607 polymorphism in mildly hypercholesterolemic adults. **J Nutr.,** v. 146, n.4, p. 720-727, apr. 2016.

WASSER, S.P. Medicinal mushrooms as a source of antitumor and immunomodulating polysaccharides. **Appl Microbiol Biotechnol.**, v. 60, n.3, p. 258274, nov. 2002.

WATERS, M.D. et al. Activity profiles of antimutagens: in vitro and in vivo data. **Mutat Res**., v.350, n.1, p.109-129.

WENG, B.B. et al. Toxicological and immunomodulatory assessments of botryosphaeran (β-glucan) produced by Botryosphaeria rhodina RCYU 30101. **Food Chem Toxicol.,** v. 49, n. 4, p. 910-916. apr. 2011.

WHEELER, H.E.; KIM, S.K. Genetics and genomics of human aging. **Philos Trans R Soc Lond B Biol Sci.**, v. 366, n. 1561,p. 43-50, jan. 2011.

WHITMAN, S.C. A practical approach to using mice in atherosclerosis research. **Clin. Biochem. Rev.**, v. 25, p. 81-93, 2004.

WHO. World Health Organization. Global Atlas on cardiovascular disease prevention and control. 2011. Available at:

<http://www.who.int/cardiovascular_diseases/publications/en>. Accessed on: May 10,

2016.

WILLCOX, B.J. et al. Midlife risk factors and healthy survival in men. **JAMA**, v. 296, p. 2343-2350, 2006.

WILLCOX, B.J.et al. FOXO3 is strongly associated with human longevity. **Proc. Natl. Acad. Sci. USA** , v.105, p.13987-13992, 2008.

WILLIAMS, D.L.; DI LUZIO, N.R. Glucan-induced modification of murine viral hepatitis. **Science,** v. 208, p. 67-69, 1980.

WILSON, D.M.; BOHR, V.A.; MCKINNON, P.J. DNA damage, DNA repair, ageing and age-related disease. **Mech Ageing Dev**., v. 129, n. 7-8, p. 349-52, jul.-aug. 2008.

XAVIER, H.T. et al. V Diretriz Brasileira de Dislipidemias e Prevençao da Aterosclerose. Arquivos Brasileiros de Cardiologia,v.101, n.4, sup.1, Oct. 2013.

XIAO, C. et al. Hypolgycemic effects of Ganoderma lucidum polysaccharides in type 2 diabetic mice. **Archives of Pharmacal Res**., v.35, n. 10, p. 1791-1801, 2012.

YADOMAE, T. Structure and biological activities of fungal beta-1,3-glucans.

Yakugaku Zasshi, v. 120, p. 413-431, 2000.

YAGI, K. A simple fluorometric assay for lipoperoxide in blood plasma. **Biochem. Med**., v.15, p. 212-216, 1976.

YANG, B.K. et al. Hypolipidemic effect of an exo-biopolymer produced from submerged mycelial culture of *Auricularia polytricha* in rats. **Biotechnol Lett**., v. 24, p. 1319-1325, 2002.

ZADELAAR, S. et al. Mouse models for atherosclerosis and pharmaceutical modifiers. **Arterioscler Thromb Vasc Biol**, v. 27, n.8 , p.1706-1721, 2007.

ZELKO, I.N.; MARIANI, T.J.; FOLZ, R.J. Superoxide dismutase multigene family: a comparison of the CuZn-SOD (SOD1), Mn-SOD (SOD2), and EC-SOD (SOD3) gene structures, evolution, and expression. **Free Radic Biol Med.**, v.1, n.33, p. 337349, aug. 2002.

ZONG, A.; CAO, H.; WANG, F. Anticancer polysaccharides from natural resources:

a review of recent research. **Carbohydr Polym.**, v.6, n. 90, supp.4, p. 1395-1410, nov. 2012.

Zhai X, Zhao A, Geng L, Xu C. Fermentation characteristics and hypoglycemic activity of an exopolysaccharide produced by submerged culture of *Stropharia rugosoannulata* #2. **Ann Microbiol.**, v. 63, n. 3, p. 1013-1020, sept. 2013.

ZHONG, N.; WEISGRABER, K.H. Understanding the association of apolipoprotein E4 with

Alzheimer's disease: clues from its structure. **J BiolChem,** v. 284, p. 6027-6031, 2009.

ZHONG, L. et al. A rapid and cost-effective method for genotyping apolipoprotein E gene polymorphism. **Mol Neurodegener**, v.11, n.2, p. 1-8, 2016.

ZHU, F.; DU, B.; XU, B. A critical review on production and industrial applications of beta-glucans. Review. **Food Hydrocol.**, v.52, p. 275-288, 2016.

MIX
Papier aus verantwortungsvollen Quellen
Paper from responsible sources
FSC® C105338

Printed by Books on Demand GmbH, Norderstedt / Germany